The Ecological Vision of J.M.G. Le Clézio

Modern **F**rench **I**dentities

Edited by Jean Khalfa

Volume 147

PETER LANG
Oxford - Berlin - Bruxelles - Chennai - Lausanne - New York

Bronwen Martin

The Ecological Vision of J.M.G. Le Clézio

PETER LANG
Oxford - Berlin - Bruxelles - Chennai - Lausanne - New York

Bibliographic information published by the Deutsche Nationalbibliothek.
The German National Library lists this publication in the German National Bibliography; detailed bibliographic data is available on the Internet at http://dnb.d-nb.de.

A catalogue record for this book is available from the British Library.

Library of Congress Cataloging-in-Publication Data

Names: Martin, Bronwen, author.
Title: The ecological vision of J.M.G. Le Clézio / Bronwen Martin.
Description: Oxford ; New York : Peter Lang, [2024] | Series: Modern French identities, 1422-9005 ; Volume no. 147 | Includes bibliographical references and index.
Identifiers: LCCN 2023030440 (print) | LCCN 2023030441 (ebook) | ISBN 9781803740676 (paperback) | ISBN 9781803740683 (ebook) | ISBN 9781803740690 (epub)
Subjects: LCSH: Le Clézio, J.-M. G. (Jean-Marie Gustave), 1940---Criticism and interpretation. | Human-animal relationships in literature. | Human ecology in literature. | French literature--History and criticism.
Classification: LCC PQ2672.E25 Z749 2024 (print) | LCC PQ2672.E25 (ebook) | DDC 843/.914--dc23/eng/20230719
LC record available at https://lccn.loc.gov/2023030440
LC ebook record available at https://lccn.loc.gov/2023030441

Cover image: Bird in Flight. Andrea Reiman, photographer. CCO 1.0 Public domain. Wikimedia Commons.
Cover design by Peter Lang Group AG

ISSN 1422-9005
ISBN 978-1-80374-067-6 (print)
ISBN 978-1-80374-068-3 (ePDF)
ISBN 978-1-80374-069-0 (ePub)
DOI 10.3726/b20410

Published by Peter Lang Ltd, Oxford, United Kingdom
info@peterlang.com - www.peterlang.com

This publication has been peer reviewed.

Contents

Introduction

This book explores the ecological vision of Nobel Prize winner J. M. G. Le Clézio, focusing on the language of animals and on the human–animal relationship. It addresses Gilles Deleuze's concept of becoming-animal, of uncovering the animal within us: it is this act of transformation that emerges as a central theme of my study.[1] As Le Clézio has declared: 'Les animaux peuvent nous enseigner quelque chose.'[2] Indeed, this statement may be all the more pertinent in the face of the ecological destruction we are now encountering, calling for a radical change in the way we look at the world and in our concept of the self. A central philosophical and ecological role is accorded the imagination, posing a challenge to the fundamental premises, rational and instrumental, underpinning Western thought and Western societies. At the same time, each of the four chapters addresses the wider political and social implications of the author's poetic vision. The chosen texts are *Haï* (1971), set in Panama; 'Nos vies d'araignées' (2011), set in the countryside;[3] *Bitna, sous le ciel de Séoul* (2018), in South Korea; and *Alma* (2017), in Mauritius and Europe. These are places where Le Clézio has resided for varying periods of time. As he declared in an interview in 2014: 'J'écris pour voyager. Je ne suis pas donc un écrivain voyageur.'[4]

In a focus on interculturality or Relation, the texts display the influence of several world philosophies including shamanism, Zen Buddhism and *créolité*. For instance, the language of *The Story of Dodo* in *Alma* is

1 As will emerge, there are very close intertextual relations between Le Clézio, Gilles Deleuze and Félix Guattari.

2 Le Clézio made this statement in a television interview: 'La Grande Librairie', 6 octobre 2017.

3 References to the valley, river and mountains suggest the landscape of Mauritius.

4 <http://www.ouest-france.fr/culture/livres/le-clezio-le-prix-nobel-nest-pas-un-ecrivain-voyageur-2623964>.

shaped by the patterns of the Creole language. Indeed, close intertextual links can be discerned here with the writings of Patrick Chamoiseau and Édouard Glissant.

Importantly, I also draw attention to the interdisciplinary nature of Le Clézio's writings, to his widening of the parameters of fiction. Like Italo Calvino, Le Clézio collapses the boundaries not only between literature/ poetic expression and philosophy, but also those between fiction and science, and between fiction and the everyday.

Chapter 1: *Haï*

Haï marks a seminal point in Le Clézio's literary development and can be viewed as an introduction to the author's ecological vision, to a mode of thinking and being in the world that will underpin the three studies that follow.[5] The chapter is directly inspired by a period Le Clézio spent living amongst the Emberas and Waunanas, the Indigenous peoples of the Panamanian rainforest where he actively participated in their shamanic ceremonies. His primary aim in this chapter is to draw our attention to the contemporary relevance of Indigenous thought and practice. The text emerges as a powerful subversion of anthropocentrism, of the nature/culture dualism that underpins Western thought and social practice.

My chapter is divided into two parts mirroring the stages of a shamanic festival. In the first section, I explore the philosophical themes underpinning Amerindian cultural practice. Here I draw attention to the central role played by material energy, by *le vivant*, calling to mind Henri Bergson's concept of an intelligence or language within all living beings and within matter. In other words, the chapter collapses the oppositions between subject and object, inner and outer, thought and life that lie at the core of Western reason. Instead, the focus is on a poetic sensorial language of expression, or what might be called its magical function, the mode of communication between all forms of life.

5 J. M. G. Le Clézio, *Haï* (Geneva: Skira, 1971).

In order to broaden the scope and to enrich the reader's experience, I explore the thematic parallels between Amerindian thought and several Western philosophers whose writings have clearly influenced the composition of the text. Here, the direct impact of the phenomenological approach of Maurice Merleau-Ponty as well as his later writings on animals can be detected. Indeed, the key role Merleau-Ponty accords the body and senses in the production of meaning is a core theme of all four books. In other words, there is a fundamental rejection of representational language viewed as a tool of power over others, non-human and human. I also draw attention to the importance of earlier research conducted by the biosemiotician Jacob von Uexküll, whose concept of the Umwelt, of animal territories, has considerably widened our understanding of animal behaviour especially that of insects and birds, thereby enriching our reading of literary texts. Indeed, his writings have been a particular source of inspiration for the poet and writer Jean-Christophe Bailly. At the same time, I address the theme of human–animal equality, bringing to the fore intertextual relations with the anthropologist Philippe Descola and the philosopher Bruno Latour. The wider political and ecological implications of this approach are also developed. My study of *Haï* is supported by brief references to Le Clézio's later poetic writings, including an analysis of short passages.

In the second part of my chapter, I focus directly on the shamanic festivals, principally on the stage named 'Beka, La Fête Chantée'. Here I examine the crucial role played by sensorial language, especially that of sound in a communication with the non-human world. I explore in this context the act of metamorphosis, a poetic rendering of Gilles Deleuze's notion of 'becoming-animal'. For instance, by imitating the voice or sound of frogs, Indigenous peoples can themselves 'become' frogs. Furthermore, it is music itself that lies at the very core of the natural world (p. 52), strengthening the central ecological theme of the interrelationship of all forms of life.

At the same time, however, Le Clézio draws attention to an Indigenous awareness of the fragility of life and of the dangers posed by introspection and by egocentrism. These dangers express themselves as illness, signalling 'la nécessité de s'exprimer' (p. 41) and highlighting the crucial role of poetic language in releasing the forces of life. As we shall see later, in Le Clézio's book *Bitna, sous le ciel de Séoul*, it is the very sound of the storyteller's words that triggers a metamorphosis, prolonging the life of the mortally ill Salomé.

In the concluding section of the chapter, I address directly the theme of transcendence or the mystical: by shedding their anthropomorphic skin in these festivals, humans can communicate with the very source or ground of life. It is in this relationship to the earth and to cosmic forces that the term 'becoming-animal' acquires an enriched significance. This theme is developed in different ways in the three chapters that follow.

Importantly, throughout the essay Le Clézio engages with the wider sociopolitical significance of Amerindian thought. The text opens and closes with images of the destructive power of the contemporary Western city. In the final pages, we are warned of the very real possibility of an apocalypse and of the urgent need for continual poetic expression. This critique in *Haï* of the Western concept of civilisation, of colonialism and indeed of war itself (pp. 13, 17–18) is developed in *Bitna* in the portrayal of contemporary Seoul against the background of the conflict between North and South Korea. It emerges as a major theme of *Alma*, arguably the author's most ambitious and far-reaching work to date.

Haï can be said then to anticipate the ecological approach I shall adopt in this book, that is known as *la zoopoétique* as developed by the Proust specialist and semiotician Anne Simon. It covers the three areas addressed above – poetic expression, the philosophical and the political – and necessarily involves a close reading of the language of the texts.[6] The essay acquires a certain prophetic power in the light of the massive ecological destruction we are now witnessing, a very real threat to life on earth.[7]

6 See Anne Simon, 'Zoopoétique', *Revue des Sciences Humaines*, 328 4/2017.

7 Le Clézio dedicated his Nobel Prize to the Innuit poet Rita Mestokosho and has supported her campaign on behalf of Indigenous cultures of Canada.

Chapter 2: Insects

This chapter opens with a close reading of this short story which appears in the collection *Histoire du pied et autres fantaisies*.[8] It will be followed by a brief study of its intratextual function. The location is indefinite but references to the forest, mountains and river may suggest the countryside, possibly the landscape of Mauritius. From the very outset, Le Clézio is calling into question widely held perceptions of the spider in the West: regarded as vermin and by some as inherently evil, it is frequently the target of mass extermination. Instead, the focus is on spiders' central ecological role in maintaining life on earth, their function as guardians of 'Mother Earth'.

In the first instance, the use of the pronoun *nous* and of the present tense heightens the sense of immediacy, drawing the reader into the text in an act of identification, of a 'becoming-spider'. Instead of presenting us with power struggles, the traditional depiction of animals in the West, the focus is on patterns of communication between the different forms of life: plants, leaves, butterflies, grasshoppers. Attention is also drawn to the theme of vigilance: the spiders need to remain alert in order to protect their young against 'les carnassiers de la nuit' (p. 215). Furthermore, they are acutely aware of the fragility of life on earth and of the need to keep on spinning to ensure the cycle of life, that day follows night. Indeed, it is this relationship from earth to sky that is foregrounded. The story acquires contemporary relevance in the light of the rapid decline of insects we are witnessing and the threat this poses to the ecological equilibrium.

In the second half of the chapter, I examine in detail the language of spiders, highlighting the poetic and the sensorial, in particular the role of sound, smell and touch. Indeed, as I shall show, the use of language is distinctive and differs considerably from that of the other stories in this collection. I draw attention to the musicality of the text, to the rhythm of

8 J. M. G. Le Clézio, *Histoire du pied et autres fantaisies* (Paris: Éditions Gallimard, 2011). The date of composition of the story of the spiders was probably much earlier.

the sentences and to the repetition of patterns and words likened to the act of spinning. The text brings to the fore the beauty of the natural world, one that can no longer be taken for granted.

In the final section, I explore the fundamental intentionality underpinning the sensorial activity and movements of the spiders. I show how the text is anchored in the semiotic structure of the quest, that of story itself. Here I draw upon the works of A. J. Greimas[9] as well as upon recent research in Paris conducted by Anne Simon and Denis Bertrand. This approach marks a further challenge to the human–animal dualism, thus fuelling the concept of 'becoming-animal'.

The Intratextual Function

Our reading of 'Nos vies d'araignées' can be further enriched when read within the context of the title of the collection, that is, within a wider web of stories. Indeed, in each text the protagonist is in danger of losing her/his footing, their relationship to the earth, to the very source of life or primordial energy. I have chosen to focus on the final story in the collection grounded in an historical event: the bombing of Tripoli, Lebanon, in 2008.

'Nos vies d'araignées' and 'Personne: sur une pensée de Ludwig Wittgenstein'

As the title may suggest (Personne: sur une pensée de Ludwig Wittgenstein) this story is a poetic exploration of Ludwig Wittgenstein's concept of being and non-being. It can be seen to throw further light, both philosophical and political, on Le Clézio's works, in particular on the theme of *le vivant*, of the birth of life and of language.

9 A. J. Greimas was the founder of the Paris School of Semiotics. He was influenced by Vladimir Propp as indeed was Le Clézio.

Set in Tripoli, the narrative moves in and out from the third-person perspective of a couple in the street to the first-person account of an unborn baby or fetus that is striving to be born. It is this account that is the principal focus of my study. It differs considerably from the use of language in 'Nos vies d'araignées' or indeed in any other text of Le Clézio. It is also somewhat darker than most of the stories in the collection although there is an expression of hope at the end.

The opening pages introduce the central theme of the text, that is, the desire of the baby to be born, its longing for form or Being. The narrator speaks of a creative force that precedes birth and that lies buried in the sky. Here it is insects that represent the possibility of life, a call to being. Fleeting sensations are evoked in expressions of sound, movement and touch. There are no complete sentences: 'Un bruit d'ailes. Un flottement. Un frôlement' (p. 135). The sensorial is then associated with the advent of language, that is, with the individual sounds of a name. Unfortunately, this desire for being is soon shattered: the insects – the butterflies – fall victim to the bombs. Any relationship to the cosmic is out of the question: the unborn child has been rendered 'aveugle et insensible'. This pattern is repeated at several intervals in the text in a focus on multiple deaths.[10]

However, the story, like that of the spiders, ends with an assertion of life, of the cosmic circle conveyed in the image of a baby who has survived the bombing and who is laughing as a fly alights on its body. The baby has been released as it were, from the tunnel of history, from 'le monde mécanique créé par les humains', evoked in the story 'A peu près apologue' (pp. 337–338). The text itself is an expression of hope, of the regeneration of life even in humankind's darkest hours.

10 Tragically, Lebanese citizens have been caught up in a number of violent political confrontations since 2008.

Chapter 3: *Bitna, sous le ciel de Séoul*

This chapter explores the language of birds and the human–animal relationship in this novel *Bitna, sous le ciel de Séoul*.[11] The book is composed of a network of seven interlocking stories recounted by Bitna, a student at a SKY University in Seoul to a young lady, Salomé, who, the victim of a degenerative disease, is confined to the space of her room. The book highlights the central role played by the storyteller and by the life-enhancing power of the imagination.

The principal focus of my study is the first story in the collection describing the relationship between M. Cho, a caretaker of a block of flats in the city, and his pigeons.[12] He is bringing them up so that one day they may return to their native land bringing messages of peace to a country separated by war, that is, the war between North and South Korea. Listening to the story of their emergence from their cages and of their flights over Seoul towards North Korea, Salomé is seen to recover a relationship with the cosmic and with the primordial forces, a relationship that Le Clézio views as mystical. Our attention is drawn to the act of metamorphosis, of deterritorialisation, or 'becoming-bird'. Intertextual links can be detected here with *Haï*, that is, with shamanism, the official religion of South Korea. Le Clézio brings to the fore the curative function of the imagination: Salomé does not allow herself to be crushed by the weight of her suffering.

Indeed, the primary focus here is on the language of birds, on their capacity to communicate with M. Cho, calling into question the nature/culture opposition underpinning Western cultures and the doctrine of human exceptionalism. Here Le Clézio is drawing upon a detailed biological knowledge of pigeons including their faculty for memory and a sense of the future. Indeed, the question of animal intelligence becomes the subject of debate between M. Cho and his neighbours who mock him

11 J. M. G. Le Clézio, *Bitna, sous le ciel de Séoul* (Paris: Éditions Stock, 2018).

12 M. Cho used to live in the countryside on the border with North Korea where his mother looked after pigeons.

for giving the pigeons individual names. We are never in the world of pure fantasy, an approach that for Salomé facilitates the process of imaginative identification.

Importantly, this presentation of the birds emerges against the background of a sustained critique of globalisation and of capitalism with their roots in humanism. Concerned with the market value of his property and the need to maintain it in pristine condition, the caretaker forbids M. Cho to use the lift to carry his pigeon baskets to the roof of the apartment building. As a result, he must climb a twenty-storey block of flats several times a day in order to provide fresh air for the birds.

In a concluding section, I directly address the political dimension of the 'The Story of M. Cho and the Pigeons', that is, Le Clézio's critique of war. In the story, although by the end of the 1960s the conflict between North and South Korea is officially over, there are still violent confrontations at the border. The earlier training exercises of the pigeons now enable them to fly safely over the high mountains and we subsequently learn of their successful arrival at a collective farm 'de l'autre côté de l'estuaire'.[13]

Le Clézio could be deemed here to be advocating peaceful resistance and to be promoting intercultural relations: in 2010, he created in Mauritius La Fondation pour L'Interculturel et La Paix. At the same time, however, Bitna, the storyteller, is aware of the historical facts: the conflict has not in fact ended. Her story remains a dream or an expression of hope.

Finally, it is important to note that 'The Story of M. Cho and the Pigeons' is embedded within a much wider network of stories that explore 'le temps réel', or everyday life in Seoul. Focusing on the experiences of women,[14] they bring to the fore the themes of cruelty, egotism and indifference to suffering. Indeed, these stories could be said socially to contextualise the act of 'becoming-bird', the need for a transformation of the self, for an awakening of the power of the senses and of the imagination.

13 The expression 'de l'autre côté' is repeated in the text and may also connote the space of origins.

14 Le Clézio taught French Language and Literature at the Ewha Women's University in Seoul in 2007.

Chapter 4: *The Story of Dodo*

This chapter, the longest in the present volume, is devoted to Le Clézio's novel *Alma*, set in contemporary Mauritius and France.[15] Importantly, there are two intersecting voices in the text, that of the Frenchman, Jérémie Felsen, who visits the island, and that of the Creole Dodo, the primary object of my study.

From the very outset, the text raises the question of identity. In the choice of the name Dodo, a contemporary rendering of the historical dodos, of the birds that could walk on the island,[16] Le Clézio collapses the human–animal duality underpinning Western cultures and philosophy. The story can be viewed as a particularly creative exploration of the concept of 'becoming-animal' lying at the core of the author's ecological vision.

This act of metamorphosis is conveyed in the very choice of narrative strategies. In contrast to the interwoven story of Jérémie Felsen, that of Dodo is written entirely in the present tense, that is, in the time of animals. It is also heavily influenced by the structures of the Creole language, a language of poetic sensorial expression, one that is totally/virtually devoid of abstraction. For instance, the story accords a central role to sound, to the sound of words and also to music, a core feature of the natural world.

Furthermore, Dodo shares the qualities of a bird. Homeless, and constantly on the move, in flight, he is in direct communication with the cosmic, traversing at the same time the different territories of the island, those, for example, of trees, of mountains and of graveyards. He is also frequently addressed as a bird: 'Hé-hon! Dodo, Dodo-bird! Kot to été?' (p. 23) and 'Dodo! Dodoo! (p. 256). This was the cry of the original birds who could walk on the island.

At the same time, Dodo during his flight is establishing multiple relations between the different cultures or philosophies on the island, as

15 J. M. G. Le Clézio, *Alma* (Paris: Éditions Gallimard, 2017).

16 Jérémie Felsen provides details of the birds that could walk on the island and of their slaughter during the colonial era. He is aware of his own family's complicity in their death.

indicated initially in the choice of names of people and places. Here I explore in depth the theme of interculturality or Relation, of a being-in-common. I also draw attention to the close intertextual relations with the writings of Glissant and Chamoiseau. Indeed, for Glissant the bird is the privileged image of Relation or creolisation. I also develop the central role of music.

The second section of my chapter is set primarily in Paris and focuses on the ecological and the sociopolitical. In the first instance, a very real threat is posed to Dodo's bird identity by the environmental destruction he encounters in Paris, in particular by the pollution of water, the very source of life. Any sensorial identification with the earth or with the cosmic is now out of the question. He returns later in the text to the landscape of Paris in a process of expansion and intensification (pp. 251–252). Here sensorial aggression is conveyed in terms of height, quantity and sound (p. 252). These renderings of the contemporary city can be interpreted as a sustained critique of capitalism.

Finally, in the image of homeless child refugees following Dodo around the streets of the city and later in that of Béchir, an Algerian war-traumatised victim, Le Clézio presents the reader with his most powerful and moving condemnation of war to date. The text highlights the need for constant poetic expression in a world enslaved to the divisive powers of reason.

However, *The Story of Dodo* ends on an uplifting note, on the curative transformative power of the imagination. In the first instance, we learn of a meeting with an artist at the port of Nice, the beauty of whose drawings transports the narrator back to the landscape of Mauritius. Later, confined to an institution, it is music itself, the singing of 'Auld Lang Syne', that becomes a mode of resurrection of life and of memory.

CHAPTER 1

Haï

This chapter examines the ecological vision of J. M. G. Le Clézio as it emerges in his poetic essay *Haï*, inspired by the period he spent living amongst the Emberas and Waunanas, the Indigenous peoples of the Panamanian rainforest. Le Clézio's text is loosely divided into three sections mirroring the stages of a shamanic ritual: 'Tahu Sa, L'Œil qui voit tout', 'Beka, La Fête Chantée' and 'Kakwahaï, Corps exorcisé'.[1] I show that in a prolonged attack on anthropocentrism,[2] on the separation between nature and culture that continues to underpin Western societies, Le Clézio argues for what the anthropologist Philippe Descola has termed 'une écologie des relations'.[3] In other words, I propose that, in a dismantling of the boundaries between human and non-human and of their roots in classical humanism, Le Clezio is presenting us with an alternative mode of perception and of being in the world; it is one which collapses the oppositions between mind and matter, language and life, the sacred and the profane as well as those between science, philosophy and the poetic imagination. It will be seen to throw invaluable light on the author's later work, providing a philosophical

1 J. M. G. Le Clézio, *Haï* (Genève: Éditions d'Art Albert Skira, 1971). In the spirit of shamanism, *Haï* is not a conventional essay but rather a series of 'explosions de language' (p. 41).

2 For the relationship between anthropocentrism and Eurocentrism, see Emmanuelle Pouydebot, *L'Intelligence animale* (Paris: Odile Jacob, 2017), pp. 154–164.

3 Philippe Descola, *L'Écologie des autres: l'anthropologie et la question de la nature* (Versailles: Quæ, 2011), p. 100. Earlier, Michel Foucault wrote: 'Le problème général de toute ethnologie est bien celui des rapports (de continuité ou de discontinuité) entre la nature et la culture'. See *Les Mots et les choses: une archéologie des sciences humaines* (Paris: Gallimard, 1966), p. 389.

and sociopolitical background to his presentation of both human and non-human animals.

Importantly, the opening sentence of the essay conveys the enormous physical impact of Le Clézio's encounter with Amerindian culture, the discovery of a hidden dimension of the self and of the world: 'Je ne sais pas trop comment cela est possible, mais c'est ainsi: je suis un Indien. Je ne le savais pas avant d'avoir rencontré les Indiens au Mexique, au Panama' (p. 5).[4] The author draws our immediate attention to the wider social and political significance of this encounter, describing it as 'une nécessité pour qui veut comprendre ce qui se passe dans le monde moderne' (p. 11). And later, in a third section, Le Clézio reminds us that 'les formes et les voix indiennes […] veulent nous dire quelque chose, et nous dévrions bien l'écouter' (p. 135).[5]

Indeed, as I shall argue, the essay calls into question the European concept of civilisation or culture. Addressed primarily to the Western reader, it emerges as a sustained critique of the privileged status accorded the human within European thought and societies:[6] 'Ils disent: la fin de l'homme est dans l'homme, les raisons du langage sont le langage même' (p. 13). Le Clézio will present us with images of the modern Western or globalised city evoked as a prison and as essentially life-destroying: 'Prisons qui enserrent les hommes et les asphyxient' (p. 138) and 'Les villes aux murs

4 In a later essay, Le Clézio describes his earlier encounter with Indigenous peoples as 'une expérience qui a changé toute ma vie, mes idées sur le monde et sur l'art, ma façon d'être avec les autres, de marcher, de manger, de dormir, d'aimer, et jusqu'à mes rêves'. See *La fête chantée et autres essais de thème amérindien* (Paris: Gallimard, 1997). Document de couverture.

5 Jean-Philippe Pierron discusses the ethical dimension of Le Clézio's poetic writing in *Les Puissances de l'imagination* (Paris: Les Éditions du Cerf, 2012), pp. 89–93. Referring to the Nobel citation, he writes: 'C'est cette capacité qu'a la littérature de questionner un moment de la culture, de nous rendre présent à notre présent – la civilisation régnante – qu'a reconnu le comité Nobel' (p. 89).

6 The presentation of Amerindian culture against the background of a critique of Western civilisation and of humanism resonates with the thought of Claude Lévi-Strauss who likewise condemns the separation of nature and culture and the Western notion of progress. See Claude Lévi-Strauss, *Structural Anthropology 2* (Penguin Books, 1994), p. 336.

infranchissables' (p. 138). Furthermore, the city symbolises the stranglehold of reason: 'Les villes ne nous laissent pas le temps de savoir. Elles ont creusé leurs pièges, elles ont tissé leur entrelacs de cause à d'effets' (pp. 13–14). It is also associated with the Western values of historical progress and political hegemony, 'la conquête de l'univers' (p. 13),[7] themes that are developed in Chapters 3 and 4 of this book in the context of war. Indeed, Le Clézio expresses in his introductory pages a deep sense of loss. Humans can no longer understand the language of the non-human world: 'L'homme ne comprend plus les animaux, il n'entend plus le langage des pierres' (pp. 30–31).[8]

I begin by exploring the central philosophical themes underpinning Indigenous culture of the Americas before going on to focus on the role of the shamanic festivals themselves. Stress is laid throughout on the modernity of Indigenous thought and cultural practice, bringing to the fore intertextual and intercultural resonances. Indeed, Le Clézio's essay will be seen to anticipate recent developments in the fields of ethology, the sciences, anthropology and philosophy as well as to make a specific contribution to ecopoetics and ecopolitics. Particular attention is paid then to the interdisciplinary nature of Indigenous thought: there is no clear boundary between the sciences (including cosmology), philosophy, poetic expression and the sociopolitical. Indeed, the essay can be said to fuel the concept of 'becoming-animal', a core theme of this book.

7 In his literary texts, Le Clézio establishes a parallel between humanism, colonial oppression and slavery. See, for example, *Révolutions* (Paris: Éditions Gallimard, 2003).

8 There are echoes here of the biblical notion of Paradise. See also Le Clézio's later novel *Le Chercheur d'or* (Paris: Éditions Gallimard, 1985), pp. 14, 23, 29. Moreover, in his short story 'Chercher l'aventure', Le Clézio writes: 'En vérité, c'est si difficile d'entrer dans le monde adulte quand toutes les routes conduisent aux mêmes frontières, quand le ciel est si lointain, […] que les animaux ne parlent plus et que les hommes eux-mêmes ont perdu leurs signes'. See J. M. G. Le Clézio, *Cœur brûle et autres romances* (Paris: Éditions Gallimard, 2000), p. 87.

Energy

As Le Clézio informs us, the title *Haï* denotes 'l'activité, l'énergie' in the native language (p. 143),[9] a reference to the Aztec metaphysical belief in the world as energy-in-motion and as a permanent oscillation between two poles.

Everything is in the process of becoming.[10] Material energy is the ground of all reality, human and non-human and it is sacred. Indeed, as James Massie has pointed out, the organising principle of Aztec metaphysics is the web evoking the interconnectedness of all forms of life, a concept that has resurfaced in ecological thought to describe what William Rueckert has termed the First Law of Ecology.[11] As Le Clézio explains in a later interview, for the Indigenous peoples of Panama, nature is not opposed to the human world, 'mais serait la nature dans le sens général d'un monde dont l'homme n'est qu'une partie, une parcelle'. What is important,

9 Referring to the human voice, Le Clézio states: 'Jamais il n'y a eu tant de vie en un seul homme, jamais tant d'énergie' (p. 91) and 'Mots magiques, dessins magiques qui étaient l'énergie de la vie' (p. 141). For the value of energy in contemporary non-Western cultures, see François Laplantine, *Penser le sensible* (Paris: Pocket Agora, 2018), pp. 9–66.

10 Teol has dual aspects life–death, being–non-being, order–disorder. It is a dynamic self-generating force, an expression of creative energy. See James Maffie, *Aztec Philosophy: Understanding a World in Motion* (Boulder: University Press of Colorado, 2013) and also his article 'Weaving the Aztec Cosmos: The Metaphysics of the Fifth Era', <https://www.mexicolore.co.uk/aztecs/home/aztec-philosophy>, 17 March 2013.

11 William Rueckert, 'Literature and Ecology', in *The Ecocriticism Reader*, eds Cheryll Glotfelty and Harold Fromm (Athens: University of Georgia Press, 1996), pp. 105–123 (p. 108). The concept of the interrelatedness of all forms of life is also central to the Norwegian philosopher Arne Naess, who in the 1970s invented the term 'deep ecology'. Donna Haraway would prefer the term 'entanglement'. As she writes: 'Everything is connected to something which is connected to something else.' See Donna Haraway, *Staying with the Trouble* (Durham and London: Duke University Press, 2016), p. 173.

'c'est le rapport de l'être humain avec ce que j'ai appelé la nature, c'est-à-dire à cet ensemble cohérent'.[12]

Indeed, as I shall show, it is energy that emerges as the dominant theme of the text, appearing in close association with the recurrent term *le vivant*.[13] Importantly, the living is presented not as a state or substance but as a creative act, an event. The opening pages draw our immediate attention to the relationship of humans to cosmic forces, that is, to creation itself: 'Là, un jour, sans qu'on sache bien comment, on se trouve en présence des raisons premières, et l'on voit ce qui pourrait venir' (p. 12). Le Clézio speaks of 'ce pays si terriblement vaste, si vivant que les choses les plus inertes battent comme des cœurs et vibrent comme des cerveaux, pays dépeuplé, pays légendaire, qui est en train de naître' (p. 12). We may be reminded here of Giorgio Agamben's concept of the origin as pure fire[14] or of Walter Benjamin's image of the whirlwind.[15] This overwhelming energy is conveyed in the cosmic image of the sea, highlighting the self's enclosure within nature and identified with the power of thought, with poetic expression: 'C'était elle (la mer) qui entourait les villes, elle qui organisait les pensées des hommes, qui réglait ses musiques, ses tableaux et ses poèmes. Et non pas l'inverse' (p. 13). From the very outset, Le Clézio is dismantling the traditional boundaries between human and non-human, between thought and life.

Indeed, this notion of energy as alive and as a creative and intelligent force calls to mind the vitalism of Henri Bergson, that is, his concept of an 'élan vital', of an intelligence or consciousness within living organisms and

12 Justyna Gambert, 'Entretien avec J. M. G. Le Clézio', *Les Cahiers J.-M. G. Le Clézio* no. 8/2015, p. 67.

13 See pp. 40, 100, 121, 125. For a detailed discussion of this concept, see *Le moment du vivant*, dir. Arnaud François et Frédéric Worms (Paris: Presses Universitaires de France, 2016), pp. 27–43. Interestingly, in his later works Jacques Derrida came close to Bergson in his understanding of the concept of *le vivant*.

14 Giorgio Agamben, *Le feu et le récit*, trans. Martin Rueff (Paris: Éditions Payot et Rivages, 2015), pp. 7–17.

15 In *Le Feu et le récit*, Agamben states that for Benjamin 'le tourbillon de l'origine reste présent jusqu'à la fin et accompagne en silence chaque instant de notre existence' (p. 72). Agamben makes it clear that 'le sujet ne doit pas être conçu comme une substance, mais comme un tourbillon dans le flux de l'être' (p. 73).

within matter.[16] Recent years have witnessed a revival of interest in his writings, in particular in his concept of *le vivant* both in philosophy and in the movement 'New Materialisms'.[17] For instance, in her book *Vibrant Matter: A Political Ecology of Things*, the materialist philosopher Jane Bennett speaks of a New Vitalism inspired by Bergson.[18] Importantly, Bergson's thought has directly influenced that of the phenomenologist Maurice Merleau-Ponty, whose text *La Nature* has played a seminal role in the development of animal studies in France.[19] It has contributed to a new way in reading texts, to 'La sémiotique du vivant' focusing on semiotic forms 'qui font vaciller la frontière entre nature et culture'.[20] This notion of *le vivant* will be crucial to the understanding of the texts examined in the present book.

The Phenomenological Perspective: Language/Thought and Life

From the very outset, the essay presents itself as a sustained critique of philosophical dualism. Indeed, it can be regarded as a phenomenological

16 See Henri Bergson, *La Conscience et la vie* (1911), dir. Frédéric Worms (Paris: Presses Universitaires de France, 2011). All forms of life are conscious, i.e. they have memory and a sense of the future. See also 'L'acte du vivant', by Rocco Ronchi, in *Le moment du vivant,* pp. 27–43.

17 'New Materialisms' is the name given to a series of movements in several fields, all of which critique anthropocentrism. It would include the life sciences, anthropology, biology and quantum mechanics. See, for example, Donna J. Haraway, *When Species Meet* (Minneapolis: University of Minnesota Press, 2008).

18 Jane Bennett, *Vibrant Matter: A Political Ecology of Things* (Durham, NC: Duke University Press, 2010), pp. 47–69. Drawing upon the theories of the biosemiotician Réné Thom, Bennett posits a shared identity between the biological processes in cells and the production of language. See Chapter 2 of the above for a discussion of Thom's catastrophe theory.

19 Maurice Merleau-Ponty, *La Nature: notes, cours du Collège de France*, ed. Dominique Séglard (Paris: Seuil, 1994).

20 Jacques Fontanille, *Formes de vie* (Liège: Presses Universitaires de Liège, 2015), p. 14. See 'Le Vivant et La Sémiotique', *Signata*, 2011, pp. 195–220.

attack on the humanist concept of nature. In the introductory pages of the first section ('Tahu Sa: L'Œil qui voit tout'), Le Clézio argues that, for the Indigenous peoples of Panama, nature is not a spectacle, nor is it simply an object of delight or a setting for human action. Rather than being eternally 'en face', as Rainer Maria Rilke[21] would put it, 'l'expérience des hommes est incluse dans l'expérience de l'univers' (p. 13). Like Merleau-Ponty[22] in *Le Visible et l'invisible*, Le Clézio dissolves the opposition between subject and object, between inner and outer: 'On est là, sur la plage devant la mer, et au même instant, on est dans la plage et dans la mer' (p. 58).[23] This relationship is evoked in the image of the skin developed later in the text: 'Au centre du corps est le feu ardent de l'existence, dont le rayonnement illumine doucement l'enveloppe la peau' (p. 122). The significance of body paintings thus emerges: 'L'Indien peint son corps parce que c'est la plus grande expérience de conscience jamais imaginée par l'homme. Il sait qu'il est vivant, il le sait. La peau est le spectacle de la vie, offert aux yeux des autres' (p. 122). We note here the crucial role played by the sense of touch in our relationship with the world.[24] Furthermore, Le Clézio stipulates that the fusion between inner and outer is also that between thought and life: 'Car, ce qu'on appelait l'intérieur, ce qu'on appelait la pensée, n'était que l'extérieur du monde'

21 See Maurice Blanchot, *L'Espace littéraire* (Paris: Éditions Gallimard, 1955). Blanchot quotes from Rilke's Eighth Elegy: 'Cela s'appelle destin: être en face et rien que cela et toujours en face' (p. 172).

22 For an illustration of the phenomenological approach of Le Clézio and of a language of animals, see Natalia Nielipowiez's article 'Quelques Traces de l'Écophénoménologie dans *Le Chercheur d'Or*', *Les Cahiers J.-M.G. Le Clézio* no 10/2017, pp. 57–67. The essay opens with a focus on a Western philosophical movement that emerged in the 1980s, one that highlights the central role of the senses in our relationship to the natural world, exploring at the same time a language of animals. It also addresses the theme of a universal intelligence.

23 Maurice Merleau-Ponty, *Le visible et l'invisible* (Paris: Éditions Gallimard, 1974). The text calls to mind Martin Heidegger's notion of being as 'Pure Presence'.

24 See also Jacques Fontanille, 'Modes du sensible et syntaxe figurative', *Nouveaux actes semiotiques,* 61–63 (1999), 1–68 (pp. 30–32). Referring to the work of Herman Parret, Fontanille reminds us that touch is 'le noyau commun de toute sensibilité [...]. Car le prédicat de base du toucher, c'est celui de *la présence pure*' (p. 30).

(p. 19) and 'les chemins de la pensée sont apparents, ce sont les mêmes que ceux de la vie terrestre' (p. 20).

Indeed, this blurring of lines between inner and outer, between thought and life, becomes a distinctive feature of Le Clézio's writings. In several novels and short stories, key actantial roles are accorded the elements, collapsing the barriers between description and narrative. Here the protagonists may seek through a process of metamorphosis to fuse with the cosmic, with the space of origins, in an act of rebirth of the self and of the world. The experience is one of pure intensity, of pure presence. A case in point is the short story 'La Montagne du Dieu vivant' set in New Mexico, where light emerges as both sender and object of the quest of the protagonist, Jon.[25] It is light 'qui l'avait conduit jusqu'à la montagne' (p. 114). Le Clézio speaks of 'l'étrange regard qui l'entourait' and of 'la présence inconnue' which 'pesait sur sa tête, sur ses épaules, sur tout son corps' (p. 117). The text foregrounds images of the skin and of penetration together with notations of sensorial intensity, in particular those of heat: 'Jon sentit qu'elle (la lumière) entrait en lui par toute la peau de son corps et de son visage. Elle brûlait et pénétrait les pores comme un liquide chaud, elle imprégnait ses habits et ses cheveux' (p. 116).[26]

Furthermore, the very title of the story, 'La Montagne du Dieu vivant', together with the reiteration of the epithets *étrange* and *inconnue* would suggest that the ecstatic experience is also an encounter with the sacred.[27] In an interview with Justyna Gambert, Le Clézio describes his own writing in terms of 'une mystique liée à la perception d'un schéma général du cosmos

25 See 'La Montagne du Dieu vivant', in *Mondo et autres histoires* (Paris: Éditions Gallimard, 1978), pp. 113–133. For a detailed analysis of the significance of light in the works of Le Clézio, see the article by Christophe Premat, 'Retrouver les voix autochtones: l'hommage de Jean-Marie Gustave Le Clézio à Rita Mestokosho', *Les Cahiers J.-M.G.Le Clézio* no 7/2014, pp. 119–133.

26 For a focus on the body and the senses in Le Clézio's work, see 'Les Vibrations de la Conscience Minéralisée dans *L'inconnu sur la terre*', by Christophe Premat, *Les Cahiers J.-M.G. Le Clézio* no 12/2019, pp. 45–55.

27 Interestingly, in the closing pages of his final book, the semiotician Algirdas-Julien Greimas, also likens the ecstatic fusion of subject and object (la saisie esthétique) to an experience of the sacred. See Algirdas-Julien Greimas, *De l'imperfection* (Perigueux: Fanlac, 1987), pp. 95, 99.

[...] Et c'est vrai que pour moi, il est difficile de concevoir que la littérature soit autre chose qu'une tentative d'approche de la nature.'[28]

Importantly, in this denunciation of the nature/culture divide, Le Clézio draws our attention in *Haï* to the themes of language and the sign: 'La véritable chair, l'intérieur de la chambre, le langage et l'âme, c'était cette réalité pleine de signes incompréhensibles, avec toutes ses fleurs, ses feuilles, ses fruits, ses peaux, ses cailloux et ses traces de pas' (p. 19) and 'le regard n'est rien d'autre que la lecture des signes' (p. 28). Indeed, Indigenous artistic expressions, be they those of basketwork, music or painting, also function as signs, as a means of communication with other life forms (p. 102). Interestingly, for the zoosemiotician Thomas Sebeok, the reading of signs and the process of message exchange is the very basis of life on earth.[29] It is this critique of representational language that lies at the core of the book and of Le Clézio's ecological vision.

Universal Intelligence

In a development of this theme of language, Le Clézio introduces the concept of a universal intelligence, the central theme of the essay and a direct attack on European humanism:

28 'Entretien avec J. M. G. Le Clézio', *Les Cahiers J.-M.G. Le Clézio* no. 8, p. 67. A further interview conducted by Lu Zhang would suggest affinities between Amerindian thought and oriental philosophies, in particular with Taoism and Zen Buddhism (pp. 87–102).

29 Thomas Sebeok (1920–2001) is the founder of zoosemiotics, the study of the language of animals based on the thought of Uxeküll. In an interview in 1995, Le Clézio states that the aim of his writing is to 'produire et non pas représenter'. See J. M. G. Le Clézio, *Ailleurs: Entretiens avec Jean-Louis Ezine* (Paris: Arlea, 1995), p. 30. For Le Clézio, as indeed for Jean-Luc Nancy, representation is a form of closure. As Nancy writes, 'Presence is born, and does not cease being born.' See Jean-Luc Nancy, *The Birth to Presence*, trans. Brian Holmes and others (Stanford: Stanford University Press, 1993), p. 2.

> Le monde n'est pas muet, n'est pas aveugle. Il a des voix pour parler, des voix qui chantent les noms et les verbes [...]. C'est sans doute la seule vérité imaginable: le monde est, tout entier, sans exception, INTELLIGENT. (pp. 102–103)[30]

There are echoes here of the thought of Benedict de Spinoza for whom there exists only one substance expressed as matter and thought, and whose writings are enjoying a contemporary revival.[31] Indeed, Spinoza posits one unified interconnected system; all finite beings are beings-in-relation. The text also brings to mind Arthur Schopenhauer's concept of a will or desire underpinning all creation and of life as essentially goal orientated.[32] Indeed, for Schopenhauer, 'les animaux sont principalement et essentiellement la même chose que nous'.[33]

Importantly, this concept of a universal intelligence or language has been given fresh credence by the anthropologist Philippe Descola, who like Le Clézio spent some time (1976–1979) living amongst an Indigenous community, in this case the Achuar of the Upper Amazon. It was this concrete lived experience that was to play a seminal role in his critique of the nature/culture divide. Indeed, as Descola informs us later in his book *Par-delà nature et culture*,[34] Indigenous peoples of the Americas believe that animals possess an interiority similar to ours and that they are able to communicate with other animals including humans, themes that will also be developed in the following chapters of the present book.[35]

30 See also *La Fête chantée*: 'Ce que je découvrais ainsi, c'était l'intelligence de l'univers, son évidence, sa sensibilité', p. 15.

31 Benedict de Spinoza, *Ethics* (London: Penguin, 1996), pp. 112–113.

32 Arthur Schopenhauer, *The World as Will and Representation*, trans. E. F. J. Payne (New York: Dover, 1969). A parallel can be drawn between his metaphysics of the will and the Amerindian concept of energy. Le Clézio makes a brief reference to Schopenhauer's essay 'Misère de la littérature', in *Histoire du pied et autres fantaisies*, p. 334.

33 Arthur Schopenhauer, *Parega et Paralipomena: Short Philosophical Essays* (The 1851 Cambridge Edition of the Works of Schopenhauer).

34 Philippe Descola, *Par-delà nature et culture* (Paris: Éditions Gallimard, 2005), p. 21.

35 For Charles Peirce, the entire cosmos participates in semiosis, that is, in the emitting of signs. Thought 'appears in the work of bees, of crystals and throughout the purely physical world'. See Charles S. Peirce, 'Prolegomena to an Apology for Pragmatism', in *Collected Papers of Charles Sanders Peirce*, eds Charles Hartshorne and Paul Weiss

It is then this concept of language as an attribute of the entire living world that emerges as central to our understanding of *Haï* and of Le Clézio's subsequent texts. And, as it becomes abundantly clear in the second section, 'Beka: La fête chantée', this language has its source in the body and the senses, presenting itself as an activity, as the expression of energy, of *le vivant*.

Indeed, Le Clézio's entire essay can be viewed as an exploration of the creative, essentially poetic role of language, what he describes as its magical function: 'Le langage est une magie. C'est-à-dire un pacte associant l'homme et l'univers' (p. 29). Everyday language, the language of the community does not differ essentially from artistic expression, in other words, from the language of art: 'TOUT EST ART' (p. 45). Indeed, Le Clézio argues the conventional notion of beauty is viewed as elitist and too close to mimesis among the Indigenous community in Panama. And, as he explains, it is this magical, poetic language that is shared by both human and animal and that makes it possible for humans to communicate with the non-human: 'Les forces magiques ne sont pas le privilège de la seule espèce humaine. Les jaguars et les pécaris, les iguanes, les crapauds, les singes, les oiseaux et les insectes sont soumis aux même lois' (p. 100).

Significantly, this notion of creativity within the animal world is supported by earlier research undertaken by the German biosemiotician Jakob von Uexküll, whose theories are enjoying a contemporary revival, contributing to our understanding of patterns of behaviour in literary texts. In his book *Milieu animal et milieu humain*, he describes how the tick constructs and interprets its environment, or Umwelt, through its senses and movements. He posits the notion not of a single world but of a plurality of different perceptual words that intersect and interact.[36] And in his discussion of animal thought, the biologist and poet Jean-Christophe Bailly writes that 'le monde n'est rien d'autre que l'interpénétration de tous ces territoires entre eux'.[37]

(Cambridge, MA: Harvard University Press, 1931–1958, Vol. I (1933), pp. 530–572 [para. 55].

36 Jacob von Uexküll, *Milieu animal et milieu humain*, 1956 (Paris: Éditions Payot et Rivages, 2010). The preface is by Dominique Lestel.

37 Jean-Christophe Bailly, *Le versant animal* (Paris: Bayard, 2007), p. 99.

Importantly, Uexküll also draws our attention to the musical nature of this relationship,[38] a theme that is directly addressed in the second section of *Haï*, 'La Fête chantée', and illustrated in the following chapters of this book.

Indeed, for Le Clézio, it is music that lies at the very core of the natural world:

'Musique omniprésente, permanente: elle habite dans les roseaux et dans la forêt' (p. 52). This point will be explored later by Gilles Deleuze and Félix Guattari in their discussion of Uexküll's thought and of animal territory in *Mille Plateaux* (1980).[39] Here they stress the importance of rhythm and of refrain, citing examples of smaller animals such as birds, spiders and wasps. And later, in *Qu'est-ce que la philosophie?*, they compare the animal with the creative artist, fuelling the concept of 'becoming animal': 'L'art commence peut-être avec l'animal, du moins avec l'animal qui taille un territoire' and 'le territoire implique l'émergence de qualités sensibles pures, *sensibilia*, qui cessent d'être uniquement fonctionnelles et deviennent des traits d'expression, rendant possible une transformation des fonctions'.[40]

Finally, and most importantly, the portrayal of animals in *Haï* acquires contemporary relevance when read in the light of Merleau-Ponty's later work *La Nature*, where he develops the findings of Uexküll. Here, in an attack on the language of representation, Merleau-Ponty draws attention to the expressive and signifying function of the animal form: 'The study of the appearance of animals takes on interest when we understand this appearance as a language. We must grasp the mystery of life in the way that animals show themselves to each other.'[41] In other words, their mere physical presence, that is, their colour movements, sounds and gestures,

38 See Uexküll, 'Toute organisme est une mélodie qui se chante elle même.' He refers to 'les différentes tonalités actantielles' (p. 158) and to 'des mélodies d'impulsions rythmiquement articulées' (p. 38). Uexküll draws particular attention in this context to the image of frogs (pp. 140–143), as does Le Clézio in *Haï*. See pp. 53–69.

39 Gilles Deleuze and Félix Guattari, *Mille Plateaux* (Paris: Éditions de Minuit, 1980), pp. 381–433.

40 Gilles Deleuze and Félix Guattari, *Qu'est-ce que la philosophie?* (Paris: Éditions de Minuit, 1991), p. 174.

41 Merleau-Ponty, *Nature: Course Notes from Le Collège de France*, ed. Dominique Séglard, trans. Robert Vallier (Evanston, IL: Northwestern University Press, 2003), p. 188.

are an expression of their thought, it is their language, their style. Indeed, the influence of Merleau-Ponty on Le Clézio's essay is pronounced. As Le Clézio writes:

> Comme le jaguar porte sa pensée dans les dessins de son pelage, comme l'agouti, le pécari, comme l'iguane porte les raies de son mystère, au-dehors, cercles jaunes et noirs de sa vie et de son âme enfin visibles, [...] comme chaque cri de chaque oiseau, chaque odeur, chaque saveur, l'Indien a décidé de porter la peau de sa pensée. (p. 130)

Equality: Human and Non-human in *Haï*

Like earlier anthropologists, Jean-Jacques Rousseau[42] and Claude Lévi Strauss, Le Clézio draws the reader's attention in *Haï* to the wider socio-political dimension of his findings. The Indigenous society of Panama is rooted in an ethics of equality not only between humans but between all life forms: 'L'homme est vivant sur la terre, à l'égal des fourmis et des plantes, il ne s'est pas exilé de son territoire' (p. 100), and earlier, 'Les chemins de la pensée sont apparents, ce sont les mêmes que ceux de la vie terrestre. Il n'y a plus de priorité' (p. 20). Moreover, all living beings enjoy the status of a person: 'Il n'y a pas d'animaux. Ni de plantes; il n'y a que des hommes, plus ou moins masqués' (p. 102).

Indeed, Le Clézio's essay can be viewed as a radical denigration of human exceptionalism. He insists that 'l'homme n'a pas triomphé. Sa pensée est un acte naturel, enlacé aux autres forces naturelles. Sa penseé ne lui donne droit à aucune victoire définitive, à aucune certitude' (p. 102). Le Clézio is clearly attacking the widespread belief in the West in the intrinsic superiority of the human, in the right to dominate and control other life

42 See Jean-Jacques Rousseau, *Discours sur l'origine de l'inégalité parmi les hommes*, ed. Jean Starobinski (Paris: Gallimard, 1969) : 'Si nous suivons le progrès de inégalité [...] nous trouverons que l'établissement de la Loi et du Droit de propriété fut son premier terme' (p. 117).

forms, a belief that is fostered both by Cartesian rationalism and by biblical discourse.[43] As Le Clézio writes in reference to Indigenous peoples:

> Ils ne veulent pas conquérir le monde [...]. Ils ne veulent pas dominer avec leurs mots, avec leurs voix. Instinctivement, l'homme élimine tout ce qui le sépare, tout ce qui le rend supérieur[...]. Il est d'emblée à l'intérieur du monde, au centre de la vie. (p. 136)

This notion of equality between human and animal – indeed between all life forms – resurfaces later in the work of Descola who, as previously mentioned, spent three years (1976–1979) living amongst the Achuar tribe of the Amazon. As Descola explains in a later interview, the discovery that the tribe accorded the same status to animals as to humans and that it was possible to communicate with them through magical ceremonies had an enormous impact on him, determining the future direction of his research:

> C'est à cette occasion que je me suis rendu compte que les rapports que les Achuar entretenaient avec les plantes et les animaux étaient tout à fait particuliers puisqu'ils les traitaient comme des personnes. Et alors que ce genre de choses ne m'était pas inconnu [...] cela m'a tellement surpris, tellement bouleversé sur le terrain que je n'ai cessé par la suite de consacrer mon énergie à essayer de comprendre cette attitude si étonnante, pour nous à tout le moins. C'est ce qui m'a amené à ce que je fais actuellement.[44]

Importantly, in the radical subversion of the nature/human divide in *Haï*, Le Clézio is presenting us with a wider concept of community, one that carries with it a number of political implications. Indeed, the Indigenous peoples of Panama condemn the notion of private property: 'Les propriétés privées sont violées, elles n'avaient rien à préserver' (p. 14). And as Le Clézio later comments: 'Les sociétés amérindiennes considéraient la possession de la terre comme une hérésie, comme quelque chose d'absurde.'[45] He posits instead the concept of a shared space, of the earth itself as a community.[46]

43 As Milan Kundera explains: 'Of course, Genesis was written by a man, not a horse.' See *The Unbearable Lightness of Being*, trans. Michael Henry Heim (London: Faber and Faber, 2016), p. 286.

44 Aliocha Wald, 'Entretien avec Philippe Descola', in *Panorama de la pensée d'aujourd'hui* (Paris: Fayard, 2016), pp. 712–734 (p. 720).

45 Le Clézio, *Ailleurs: Entretiens avec Jean-Louis Ezine*, p. 58.

46 See Rousseau, *Discours sur l'origine de l'inégalité parmi les hommes*, p. 101.

The essay resonates here with the thought of Aldo Leopold, who calls for a land ethic, whereby 'the boundaries of community are enlarged to include soils, waters, plants and animals or, collectively, the land.[47] The introduction to the French version was written by Le Clézio himself. The philosopher Dominique Lestel has also spoken in this context of a hybrid community: 'Pour moi, être humain, c'est vivre dans une communauté hybride',[48] a point that is also developed in his book *Les Amis de mes amis.*[49]

These themes are later explored in Le Clézio's novel *Ourania*, set in contemporary Mexico, where he describes the utopian community of Campos prior to the seizure of homes and land by property developers, 'les puissants de la Vallée'.[50] Here the inhabitants are brought up to live in harmony with nature and the cosmos in a society where there is no distinction between thought and life. Significantly, Le Clézio's description of Campos is also influenced by his earlier discovery in Mexico of the Indigenous city Santa Fe de La Laguna, founded on the principles of Thomas More. Indeed, it is the realisation that 'l'utopie a existé, elle a été réalisée au moins une fois' which has strengthened Le Clezio's belief in the possibility of a better world and his view of literature as an utopian pursuit.[51]

In its concept of community, *Haï* can also be said to anticipate the growth of a political ecology in the 1980s and 1990s. In his book *Politiques*

47 See Aldo Leopold, *A Sand County Almanach* (New York: The Library of America, 2013), p. 116. For Leopold the value of wildness was essentially spiritual. The book was translated by Anna Gibson as *Almanach d'un comté des sables* (Paris: Flammarion, 2017). The introduction was written by Le Clézio. The text may call to mind Jean-Luc Nancy's ontological concept of 'being with', of sharing space and time with all forms of life. See *The Ecocriticism Reader* where reference is made to *A Sand County Almanach* (p. 310). Here Don Scheese suggests that for Leopold, the wilderness also had scientific value (p. 310).

48 Maria Esther Maciel, 'Interview with Dominique Lestel', *Contemporary French and Francophone Studies*, 16/5 (2012), pp. 711–719 (p. 714).

49 Dominique Lestel, *Les Amis de Mes Amis* (Paris: Éditions du Seuil, 2007).

50 See J. M. G. Le Clézio, *Ourania* (Paris: Éditions Gallimard, 2006), p. 281.

51 See Le Clézio's interview with Jacqueline Dutton, in *Le Chercheur d'or et d'ailleurs: L'Utopie de J.M.G. Le Clézio* (Paris: L'Harmattan, 2003), p. 281. This Indigenous city was considered the ideal place to put into practice the vision of Thomas More.

de la nature: Comment faire entrer les sciences en démocratie, the socialist and philosopher Bruno Latour argues for the cohabitation of the multiple forms of life and for their political representation.[52] And in his book, *Nous n'avons jamais été modernes: Essais d'anthropologie symétrique*, Latour refers to 'le parlement des choses'.[53]

Indeed, this focus on equality and on a hybrid space emerges as a central theme of several of Le Clézio's texts. By way of example, I shall briefly look at a short extract from Le Clézio's novel *Révolutions*.[54] The extract is one of the several stories that Aunt Catherine recounts to her nephew, Jean Marro, evoking her childhood experiences and those of her sister in the garden of her home in Mauritius.

The passage opens with a phatic address, a notation of surprise, of the unexpected, signalling a break with the everyday: 'Tu ne peux pas imaginer, il y avait tellement d'oiseaux, là-bas, à Ébène [...] Ils ne s'arrêtaient pas de pépier' (p. 30). The birds are introduced here in impersonal terms of pure presence ('il y avait') and of sheer energy evoked in the sensorial category of sound. The physical impact is then heightened in an emphasis on visual qualities, on differences of colour and shape, thereby drawing attention to their singularity or individuality: 'Des insolents, de jolis petits oiseaux jaunes, des bengalis, des cardinaux tout rouges, d'autres qui ont une huppe, on les appelle des condés'. We may be reminded here of Gilles Deleuze and Félix Guattari's description of animal territory as 'un jaillissement de traits, de couleurs et de sons'.[55]

After conveying the autonomy and indeed freedom of the non-human subjects, Aunt Catherine goes on to evoke patterns of friendship between the birds and the children. Significantly, this friendship is initiated by the birds themselves and expressed in a language of movement: 'Partout où nous allions, ils nous suivaient, ils volaient de branche en branche, ils s'amusaient à nous défier, ils se moquaient des chats' (p. 30). The birds are portrayed

52 Bruno Latour, *Politiques de la nature: Comment faire entrer les sciences en démocratie* (Paris: Éditions La Découverte, coll. 'Armitaire', 1999).

53 Bruno Latour, *Nous n'avons jamais été modernes: Essai d'anthropologie symétrique* (Paris: Éditions La Découverte, 1997), pp. 194–198.

54 Le Clézio, *Révolutions*, p. 30.

55 Deleuze and Guattari, *Qu'est-ce que la philosophie*, p. 174.

here as powerful actantial subjects, capable of emotion and engaging in play. Their behaviour would substantiate recent research in both the sociability of animals and their friendship with humans.[56]

At the same time, the unfolding of the passage suggests a growing awareness and understanding of the language of birds on the part of the narrator in a further narrowing of the gap between human and non-human. In a switch to the first person, we learn that Aunt Catherine is now able to imitate their songs and to open a conversation with them: 'Alors je connaissais tous les oiseaux, je pouvais imiter leurs chants et ils me répondaient' (p. 30). A dialogue is then initiated by an individual cardinal – now perched on a branch – whose song, a series of whistles, slurs and trills, is conveyed principally through the use of onomatopoeia: 'il sifflait gravement, il faisait des roulades très douces, twirr, twirr, ensuite très aigues, une sorte d'appel, il entrouvrait à peine son bec et il criait fwit, fwit, fuyiit, et moi je lui répondais' (p. 30). Aunt Catherine is clearly undergoing a transformation in this episode: she is 'becoming-bird'. Significantly, this passage would suggest a detailed biological understanding of the song of birds, a fusion of poetry and science.[57]

Indeed, the production and then imitation of the language of birds here highlights the importance of sound as a mode of communication between animal and human.[58] The choice of the term 'une sorte d'appel',

56 See Lestel, *Les Amis de mes Amis*, p. 9.

57 For the study of the language of birds and the syntactical structure of birdsong, see Jean-Claude Ameisen, *Sur les Épaules de Darwin: Retrouver l'Aube* (Paris: Éditions Les Liens Qui Libèrent, France Inter, 2014), pp. 288–289. He argues that the part of the brain where birds learn to sing is similar to that of humans when learning to speak.

58 Sound/rhythm is considered the language of creation and is a central theme of *Haï*. The importance of orality is conveyed in this passage in Le Clézio's use of language. Eschewing the use of subordinate clauses or logical connectors, he favours a process of accumulation of main clauses conveying increased sensorial intensity or energy. In a later discussion with Claude Cavallero about two of his novels, Le Clézio explains: 'Bien que n'étant moi-même musicien, j'ai eu le sentiment de concevoir ces romans à la manière de phrases et de mouvements musicaux, selon un tempo allant s'accélérant.' See 'Les Marges et l'Origine: entretien avec J.M.G. Le Clézio', *Europe* no. 954–958/janvier–février 2009, pp. 29–38 (p. 31).

suggests an invitation to join the birds in their singing in a celebration of *le vivant*, of the act of creation. As Bailly writes: 'Nous voyons que la seule universalité [...] est une universalité de l'adresse: l'univers qui est parlant se parle à lui-même et nous parle'.[59] And, as he also reminds us, it is the process of naming itself that separates us from the non-human world, from Being:[60]

> Aussitôt qu'un nom est donné quelque chose nous échappe [...]. C'est pourquoi, dans l'espace exubérant de leurs vies et de leurs parures, les animaux nous semblent toujours figurer ce qui se dérobe, et c'est pourquoi aussi dans la vérité de leurs actes et de leurs mouvements nous croyons un instant apercevoir, confondu et comme lové dans leurs apparences, cet au-delà/en deçà du langage que nous appelons la chose même, qui est ce vers quoi nous sommes ou devrions être tout entiers tendus.[61]

Indeed, for Le Clézio, birds will become the privileged image of the transcendence of the self, of the flight of the imagination. As he writes in *Haï*: 'Dépouillé de ses mots, l'homme est un cormoran. Il plane, il étend ses ailes, et il vole longtemps, longtemps, sans même bouger' (p. 90).

Significantly, the theme of equality, of partnership between human and non-human evoked in *Haï* is not only conveyed in Le Clézio's treatment of the small and familiar but also in the portrayal of wild animals, of those conventionally associated in the West with acts of extreme violence and cruelty. A case in point is the short story 'L'Arbre Yama', set in Liberia and that describes how a hyena saves the lives of two girls in flight from the forces of Charles Taylor.[62] The focus here is on expressions of altruism and empathy by the hyena, suggesting a moral status equal to that of humans. We note the role played not only by sound – the hyena's grunt – in establishing friendship with the girls but crucially that of smell in saving their lives: the hyena returns every night to leave traces of her urine by the

59 Jean-Christophe Bailly, *Le parti pris des animaux* (Paris: Christian Bourgois, 2013), p. 104.

60 This notion of Being associated with Parmenides and the pre-Socratics is also a central theme of Le Clézio's novel *Le Procès-Verbal* (Paris: Éditions Gallimard, 1963). Here it is associated with a mystical state, 'le niveau de l'ineffable' (p. 235).

61 Bailly, *Le parti pris des animaux*, p. 112.

62 'L'Arbre Yama', in *Histoire du pied et autres fantaisies*, pp. 123–153.

tree in which the girls are hiding, with the dual aim of effacing their scent and of deterring the soldiers from arriving.

The Political Dimension of Le Clézio's Writing: Colonialism and Global Capitalism

It is this wider political and historical context of Le Clézio's ecological vision that is developed in several of his fictional works where anthropocentrism and the denial of the ultimate value of life, of *le vivant*, are seen to go in hand with expressions of extreme violence. Indeed, Le Clézio can be said to present the dire consequences of the nature/culture divide and of the separation of language from life. Several of his novels explore the role of Western colonialism and global capitalism in the destruction of the world's communities, human and non-human alike. In the novel *Révolutions*, set in Nice during the Algerian war of independence, a city where 'l'abstraction régnait' (p. 214), racist murders and violent attacks on Algerians can be paralleled with the imminent destruction of 'Le Jardin des Oliviers' – a space of contemplation and communication with the cosmic – to make way for luxury high-rise apartments. Images of petrification, of heaviness and of oppression abound in the descriptions of the city. Reference is made, for example, to 'les palmiers immobiles dans leur pots' (p. 93), swallows are described as giving out 'des sifflements inquiétants, oppressants, pleins de hâte et de violence, dans le silence engourdi du dimanche après-midi' (p. 89), and the sea itself becomes 'une mer épaisse, lourde, sur laquelle les hommes flottaient comme des débris le long du rivage, parmi d'autres débris' (p. 94). Similarly, in the novel *Désert*,[63] the exploitation of immigrants in contemporary Marseille and their reduction to the status of slaves imprisoned within their ghettos, within 'les vieux murs lépreux (p. 282) and 'les ateliers étouffants' (p. 271), are mirrored in the harshness of the landscape: 'Elles [les premières

63 J. M. G. Le Clézio, *Désert* (Paris: Éditions Gallimard, 1980).

mouettes] ne ressemblent pas du tout à des princes de la mer; elles sont gris sale, avec un bec jaune et un œil qui brille durement' (p. 244).

The Role of the Shamanic Festivals

'Je crois que la nature à laquelle peut avoir accès l'humain, c'est par le biais du langage et donc d'une mise en forme.'[64]

I shall now go on to examine the sociopolitical and liberating role played by the shamanic festivals, by artistic expression within the Indigenous community of Panama. Referring to pictorial art, Le Clézio comments: 'Comme la musique, l'art est fait pour défendre, pour attaquer, pour conduire la guerre avec le monde' (p. 111). Far from presenting us with an idyllic society, 'a land of milk and honey', Le Clézio draws our attention to the fragility of the equilibrium between the forces of creation and those of destruction: 'Il n'y a pas de douce harmonie' (p. 86). Indeed, Indigenous peoples are shown not to take life on earth for granted; they are engaged in a ceaseless battle between good and evil, between spirits and demons. As the narrator explains at the very end of the text, the forces of energy, those of *Haï*, are continually coming up against those of *wandra*, of 'la submission, la domination, la possession' (p. 143).[65] Le Clézio stresses

64 See the interview conducted by Justyna Gambert, 'Entretien avec J.M.G. Le Clézio', *Les Cahiers J.-M.G. Le Clézio* no. 8, p. 6.

65 In an interview with Claude Cavallero, Le Clézio comments on the strong moral sense of the Indigenous peoples of Panama and the impact this has had on his work. He states that, although the writer should not possess any explicit moral intention, 'sa création elle-même en tant qu'acte, se charge d'une valeur morale […]. Mes livres reflètent par là ce que l'individu accepte ou refuse autour de lui; c'est une approche qui doit sans doute au rapport que j'entretiens avec le monde amérindien, dont la vision fondée sur une culture d'expérience distingue avec une grande netteté entre ce qui s'avère "bien" – entendu comme ce qui ne nuit pas aux autres et va dans le sens d'une liberté réelle – et ce qui est "mal" – qui reste donc condamnable.' See Claude Cavallero, 'Les Marges et Origine: Entretien avec J.-M.G. Le Clézio', *Europe* no. 954–958/janvier–février 2009, pp. 29–38 (p. 37).

then the key role played by the language of the festivals, the language of the community, in the creation of life and in the struggle against silence, against passive acceptance and inaction: 'Le chant indien, comme la musique indienne n'est pas seulement un prolongement du langage, il en est la défense' and 'Mots magiques, dessins magiques qui étaient l'énergie de la vie. Ils luttaient contre l'empire de la soumission' (p. 141).[66] The text calls to mind Deleuze's definition of the creative act as an 'acte de resistance' and his statement that 'résister signifie libérer une puissance de vie'.[67] Indeed, in *Haï* submission is presented both as a physical and a mental state: it is a lack of energy or powers of expression and an illness, a disturbance of the balance between life and death: 'La maladie, la folie, les dangers de mort apparaissent [...]. Ils sont les signes de la nécessité de s'exprimer' (p. 41). As Deleuze asserts in his book *Critique et Clinique*, 'La littérature apparaît alors comme une entreprise de santé' and the writer is 'plutôt médecin, médecin de soi-même et du monde'.[68]

It is, then, the language of the festivals, that of poetic expression that triggers a liberation of the self, allowing communication with the natural world: 'Signes [...] pour que les autres règnes vivants les lisent et les comprennent' (p. 102). As already noted, central importance is attached to sound in communicating with animals. Through the use of makeshift instruments, by blowing through reeds, 'le bruit devient langage' (p. 54). The experience is one of pure intensity, of pure energy: 'Par le moyen des instruments monophones, l'homme retrouve un autre langage, plus violent, plus rapide, qui le fait entendre du règne animal, végétal, ou démoniaque. Ce langage n'est plus discursif, mais répétitif' (p. 51). A process of identification takes place whereby humans become animals, recalling Deleuze's

66 In Le Clézio's novel *Poisson d'Or* (1997), it is music that enables the protagonist, Laïla, to overcome her physical and psychological oppression and to recover her space of origins. She recites passages from Aimé Césaire (*Cahier d'un retour au pays natal*) and quotes abundantly from Frantz Fanon and Vladimir Lenin in her baccalauréat (pp. 206–207).

67 See Alain Boutang, *L'Abécédaire de Gilles Deleuze et Claire Parnet*, online recording, 9 August 2013, <http://jef-safi.net/spip/spip.php?article292>. See also Gilles Deleuze–Claire Parnet: *Dialogues* (Paris: Flammarion, 1996).

68 Gilles Deleuze, *Critique et Clinique* (Paris: Éditions de Minuit, 1993), p. 14.

concept of 'becoming-animal': 'Ivresse de l'identité, la seule identité. Les hommes font de la musique avec des tubes de bambou, et ils sont des crapauds' (p. 62) and 'Les crapauds ne sont plus les crapauds, ce sont les hommes qui sont les crapauds' (p. 63).[69] Language thus becomes 'une action commune, une cohésion, un ciment' (p. 54). As the zoosemiotician Anne Simon has pointed out, it is literature, that is, poetic expression that can facilitate this process of identification.[70] She writes that 'c'est à travers des délocalisations et des déplacements internes à notre langue même que nous pouvons rejoindre l'altérité, ou lui donner lieu'.[71] A central philosophical and ecological role is thus accorded the activity of the imagination, posing a challenge to the fundamental premises, rational and instrumental, underpinning classical Western thought and Western societies.

It is not only musical instruments, however, but also the human voice that becomes a vehicle of transcendence and means of communication with the non-human: 'Le chant est plutôt une émission de sons [...] un langage pour les dieux, les animaux et les plantes, un langage pour atteindre tout ce que la parole ne peut pas atteindre' (p. 78).

Indeed, central importance is accorded in the text to the invisible: by imitating the voice of animals – of monkeys, agoutis and birds – and by shedding our anthropomorphic skin, 'l'homme ouvre un passage en lui-même. L'invisible peut entrer' (p. 80).[72] The experience is described as one of beauty, the only true expression of human freedom (p. 135). It is clearly a mystical experience:

69 See *Critique et Clinique*, p. 11: 'L'écriture est inséparable du devenir: en écrivant, on devient-femme, on devient-animal ou végétal' and 'Devenir n'est pas atteindre à une forme [identification, imitation], mais trouver la zone de voisinage, d'indiscernabilité ou d'indifférenciation.' Deleuze cites Le Clézio's writing as an example: 'Ces devenirs s'enchaînent les uns aux autres suivant une lignée particulière, comme dans un roman de Le Clézio' (p. 11).

70 See Anne Simon, 'Zoopoétique', *Revue des Sciences Humaines*, 328 4/2017.

71 See 'Entretien sur la zoopoétique avec Anne Simon', <https://www.fabula.org/colloques/document5368.php>, p. 3.

72 The transcendent or the experience of the invisible, is a recurrent theme in many of Le Clézio's texts. See, for example, *L'inconnu sur la terre* (Paris: Éditions Gallimard, 1978) and *Désert* (Paris: Éditions Gallimard, 1980).

> Il (le langage) travaille à la grande métamorphose que l'homme, au fond, ne cesse d'espérer, celle qui le libérera de sa prison anthropomorphique. Il approche de cet instant miraculeux, mythique, où l'ancêtre des hommes a créé le monde: étant l'instigateur de toutes les formes d'existence, et l'inventeur de tous les langages. (p. 51)

In other words, it is through creative expression, through this act of metamorphosis, that humans can aspire to the space of animals, can 'become animal' a central theme of the three chapters that follow. It is the space of the Open, or 'the other side'. As the poet Rilke reminds us:

> De tous ses yeux la créature voit
> L'Ouvert. Nos yeux seuls sont
> comme renversés…[73]

And Maurice Blanchot describes 'the other side' in this context as 'le fait d'être, dans ce rapport, hors de soi, dans la chose même et non dans une représentation de la chose'.[74] It is a fusional experience likened to that of ecstasy.

Interestingly, Le Clézio addresses his Nobel speech to the sorcerer and storyteller Elvira, whom he met in the Panamanian rainforest. He describes her singing as 'la poésie en action' and as the expression of the true power of nature:

> Comme si elle portait dans son chant la puissance véridique de la nature, et c'était là sans doute le grand paradoxe, que ce lieu isolé, cette forêt, la plus éloignée de la sophistication de la littérature, était l'endroit ou l'art s'exprimait avec le plus de force et d'authenticité.[75]

Significantly, towards the end of the essay, the author returns in circular fashion to the image of the modern city associated with the exclusion of *le vivant*, with imprisonment and death. In the image of the roof about to collapse, Le Clézio draws attention to the precariousness of life on earth,

73 See Rainer Maria Rilke, *La Huitième Élegie de Duino*, quoted by Maurice Blanchot in *L'espace littéraire* (Paris: Éditions Gallimard, 1995), p. 172.

74 Ibid.

75 J. M. G. Le Clézio, 'Dans la forêt des paradoxes', Conférence Nobel, 7 décembre 2008.

and to the very real possibility of an apocalypse, an issue that has resurfaced in contemporary debates:[76]

> Qui va prendre un charbon, une craie, un couteau [...] ? Qui va peindre son corps et son visage, pour maintenir encore un peu les murs de la prison, pour empêcher le plafond de descendre, pour que tout soit innocent, pour que tout le monde parle à nouveau à tout le monde ? (p. 142)

In a later essay on his experiences in the Panamanian jungle, Le Clézio returns to this theme in a celebration of the ecological conscience of the Indigenous peoples he encountered:

> Ce peuple, l'un des plus isolés et des plus démunis de la terre, [...] ces hommes ont néanmoins reçu, comme une grâce divine, l'intuition de la destruction imminente du monde, et en même temps le pouvoir d'un exorcisme, l'idée d'un salut. Ils ont la volonté et la foi de sauver tous les hommes, et pas seulement eux-mêmes.[77]

In the face of a new biblical flood and of nuclear war, he calls for creative expression – dance and music – as a weapon and act of solidarity:

> Que peut un écrivain contre le nouveau déluge qui menace le monde actuel? Que peut un homme, n'importe quel homme contre la mort thermonucléaire que la science a su inventer pour mieux se détruire elle-même ? Que peut un homme ici, dans les villes d'Europe ou d'ailleurs, pour tenter de sauver les matins du monde? Mais peut-être qu'il peut, comme les Waunanas de la forêt, simplement danser et faire sa

76 See, for example, Bruno Latour, *Face à Gaïa* (Paris: La Découverte, 2015). Latour has described our era as that of the Anthropocene, one in which humans are for the first time changing the face of nature (pp. 147–191). The conviction that a civilisation based on fossil fuels will collapse during the 2030s has given rise to a new movement in France known as 'la collapsologie'. See Cécile Bouanchaud et Audrey Garric, 'Le succès inattendu des théories de l'effrondrement', *Le Monde*, 6 février 2019, pp. 6–7. See also Timothy Clark, *The Value of Ecocriticism* (Cambridge: Cambridge University Press, 2019), where he discusses the relationship between the Anthropocene and narrative form (pp. 82–89).

77 Le Clézio, *La Fête chantée et autres essais de thème amérindien*, p. 239.

> musique, c'est à dire parler, écrire, agir, pour tenter d'unir sa prière à ces hommes et ces femmes autour de la pirogue. [...] Écrivons, dansons contre le nouveau déluge.[78]

To conclude, *Haï* can be seen to directly address the ecological function of literature. Poetic sensorial expression can become an agent of metamorphosis, possessing an essentially performative or magical function: it produces both a transformation of consciousness, a 'becoming-animal', and a regeneration of life itself.[79] At the same time, it becomes a mode of resistance against warfare, against the violence of human reason and of contemporary science. It is within this wider sociopolitical and philosophical context that the presentation of animals – non-human and human – in the three chapters that follow will acquire enriched meaning.

78 See *La Fête chantée*, pp. 239–240. In a recent book, Amitav Ghosh has promoted Indigenous cultures as a response to the climate crisis we are now witnessing. See *The Nutmeg's Curse: Parables for a Planet in Crisis* (London: John Murray, 2021).

79 See William Rueckert, 'Literature and Ecology', in *The Ecocriticism Reader*, pp. 108–111.

CHAPTER 2

Insects

> Et c'est vrai que pour moi, la littérature soit autre chose qu'une tentative d'approche de la nature. Je crois que l'être humain n'a pas la même capacité et n'a pas la même possibilité tout simplement de vivre la nature qu'ont les plantes et les animaux.[1]

This chapter opens with a study of the short story 'Nos vies d'araignées' appearing in the collection *Histoire du pied et autres fantaisies* (2011). I give a close reading of the language of spiders, addressing at the same time their wider ecological and sociopolitical function. This is followed by an analysis of the role of insects in Le Clézio's short story 'Personne', where I focus on the death of an embryo, the victim of the bombing of Tripoli, Lebanon, in 2008. The choice of texts here may illustrate the two interrelated poles of Le Clézio's writings, that of the creation of life and that of its destruction.

Spiders have for centuries occupied a privileged position in literary and philosophical writings. For Benedict de Spinoza, for instance, the image of the spider within its web is the expression of mind within the world, the doctrine of one substance. In African, Caribbean and Native cultures, the insect is frequently accorded equal status with the human.[2] And importantly, the story of Arachné, of the human who was turned into a spider, anticipates a central theme of all four books under discussion , that of metamorphosis.[3] In addition, the spider is associated by many artists

1 Quote from Le Clézio, from an interview with Justyna Gambert, *Les Cahiers J.-M.G. Le Clézio* no. 8, p. 67.

2 Interestingly, the spider is a goddess for the Navajo, the Indigenous peoples of New Mexico, whose language Le Clézio attempted to learn. See 'Les Amérindiens et nous': <https://bibliobs.nouvelobs.com>.

3 See Jean-Marc Drouin, *Philosophie de l'insecte* (Paris: Éditions du Seuil, 2014), p. 108. The origin of the story is Ovid's *Metamorphoses,* Book 5.

and writers with poetic expression and with beauty, themes that are amply illustrated in this story.[4]

'Nos vies d'araignées', a striking example of language within the natural world, evokes the everyday life of a community of spiders which, through the act of spinning and the ability to interpret their environment, are able to protect 'les petits animaux' in their burrows from predators, thereby ensuring the continuation of the cycle of life (pp. 214, 215, 216). In other words, spiders will be seen to play a key ecological role in maintaining life on earth, the story acquiring contemporary relevance in the light of their rapid decline in recent years, of their loss of habitat, the outcome of climate change and the exploitation of the land.

Significantly, the story is narrated in the first-person plural: this choice of narrative stance together with the use of the present tense heightens the sense of immediacy, drawing the reader into the text in an act of identification.[5] Le Clézio could be said to be fully exploiting the cognitive function of literature, that of introducing us to the point of view of the Other, in this case of an insect conventionally regarded in the West as vermin. The reader, like the Indigenous peoples in *Haï,* undergoes a metamorphosis, a becoming-spider in the language of Deleuze.[6] The text is reminiscent of Wajdi Mouawad's story of the fireflies, likewise related in the first-person plural and the present tense.[7] In both cases, the focus is on community, on a shared enterprise.

4 See, for example, the enormous spiders of the artist Louise Bourgeois. See also Émile Blanchard, *Les Araignées* (Paris: Éditions VillaRrose, 2003) and Dominique Jacobs, *Splendeurs de l'araignée* (Paris: Klincksieck, 2018).

5 For a study of the point of view adopted by writers in their portrayal of animals, see Sophie Milcent Lawson, 'Zoographies. Traitements linguistiques et stylistiques du point de vue animal en régime fictionnel', *Revue des sciences humaines,* 328 4/2017, pp. 91–106.

6 See *Critique et Clinique,* 'Quand Le Clézio devient Indien […] il entre dans une zone de dévoisinage plutôt qu'il n'acquiert des caractères formels', p. 12.

7 See Wajdi Mouawad, 'Lampyris Noctiluca', in *Anima,* trans. Linda Caboriau (Vancouver, BC: Talonbooks, 2017), p. 230.

The Senses

'Nos vies d'araignées' can be seen to illustrate two interrelated themes of *Haï*: *le vivant*, and the inseparability of thought from expression, from the activity of the senses. There is an explicit attack on reason and on explanatory discourse: the spiders declare that 'nous n'aimons pas expliquer les choses' (p. 213).[8] A sense of energy is conveyed in a repetition of the same sensorial categories, the same verbs.

Indeed, a key role is accorded sound in the text and the act of listening. For the spiders who live on ground level, hearing becomes a means of knowledge of the real and a mode of communication with other forms of life. The syntagm 'nous écoutons', is repeated frequently in a structure recalling that of the ritornello (pp. 211, 212, 213, 215). Bailly speaks in this context of 'cette pensivité faite avant tout d'écoute et d'attention extrêmes'.[9]

Moreover, it is a question of feeling sound, that is, of sound and touch, given that spiders have very poor hearing: 'Non, non, nous écoutons intensément, nous sentons très fort ce qui arrive' (p. 214).[10] Sensorial energy, an expression of *le vivant*, is conveyed throughout in the use of synesthesia.[11]

8 The text contrasts sharply with an earlier evocation of spiders in Le Clézio's essay *L'Extase matérielle* (Paris: Éditions Gallimard, Collection Folio/ Essais, 1967), pp. 165–168. Here the description is predominantly visual and from the perspective of a detached observer.

9 See Bailly, *Le parti pris des animaux*, p. 46.

10 Spiders, like bees, can feel sound but they are unable to hear sounds in the air. In the presentation of arachnids/insects, parallels can be drawn with the works of Samuel Beckett. See, for example, Steven Connor, 'Making Flies Mean Something', in *Beckett and Animals*, ed. Mary Bryden (Cambridge: Cambridge University Press, 2015), pp. 139–152. In both cases, the presentation of animals/insects is supported by zoological research.

11 In *La Pensée sauvage* (1962) and later in *Mythologiques* (1964–71), Lévi-Strauss develops the theme of synesthesia. He views this as a characteristic of all living beings. See Boris Wiseman, *Lévi-Strauss, Anthropology and Aesthetics* (Cambridge: Cambridge University Press, 2007), pp. 112–131. In an interview with Claude Cavallero, Le Clézio recognises the central role played by synesthesia in his writing, by 'ce métissage sensorial constitutif de notre être'. See Cavallero, 'Les Marges et L'Origine: Entretien avec Le Clézio', p. 34.

For example, sound, touch and movement come together in images of penetration, in a fusional experience: 'Les bruits, nous voulons aussi les dévorer. Ils viennent de partout [...] et ils appuient dans nos ventres au point le plus sensible' (p. 210) and 'Nous sommes sur tous les points sensibles de la terre, et chaque onde qui parcourt les rochers, les arbres, les herbes, va jusqu'à nos pattes et entre dans nos corps. C'est ainsi que nous écoutons le monde' (p. 212). Particular attention is drawn to the act of vibration in a reiteration of the verb *vibrer*: 'L'air vibre dans les toiles invisibles, vibre jusqu'à nos ventres et nous connaissons l'histoire qui se déroule' (p. 215) and 'La rivière au bruit de tonnerre use la terre et vibre et tremble mais nous n'allons jamais vers elle' (p. 212). Indeed, in this focus on spiders, that is, on an insect with very poor vision, the text also highlights the crucial role played by the sense of touch in the production of meaning, a sense that, for the semiotician Herman Parret, lies at the very core of synesthesia.[12] And in his book *Listening*, the philosopher Jean-Luc Nancy speaks of 'being as resonance' and of the 'subject as the place of resonance'.[13] Indeed, it is clearly the sense of touch that plays a central role in the construction of meaning by both humans and animals/insects. As Uexküll explains, 'l'espace tactile joue chez beaucoup d'animaux un rôle tout à fait extraordinaire'.[14]

At the same time, sensorial activity is associated with the act of creation, that is, the creation of life, with *le vivant*. The spiders, like the Indigenous peoples in *Haï*, are alert to the need for constant expression, to keep on spinning in the struggle against the forces of *wandra*, those of 'la soumission, la domination, la possession' (*Haï*, p. 143). There are multiple references here to the need to stay awake: 'Nous ne dormons pas' (p. 214), 'nous autres, nous veillons, nous restons vigilantes' (p. 214) and 'nous ne dormons pas,

12 See Hermann Parret, *Épiphanies de la présence* (Paris: Presses Universitaires de Limoges, 2006), pp. 63–68. For Parret, all sensations are based on bodily contact, that is, on touch. And for Merleau-Ponty, there exists, at a preconscious level, a basic unity of the senses. Relations between them are only possible because of this state of indifferenciation.

13 Jean-Luc Nancy, *Listening*, trans. Charlotte Mandell (New York: Fordham University Press, 2006), p. 21. Nancy also appears to draw a parallel here between animal thought and human thought. See p. 22.

14 See Uexküll, *Milieu animal et milieu humain,* pp. 124–125.

nous ne dormons jamais' (p. 215). We are reminded here of Deleuze's description of the animal as 'l'être aux aguets'.[15] And to quote Bergson: 'C'est la conscience ou mieux la supra conscience qui est à l'origine de la vie'[16] and 'cette conscience est une exigence de création'.[17]

This association between the exercise of the senses and the ability to interpret the real is foregrounded in the reiteration of the term 'savoir': 'Tout ce qui se passe à ras de terre, nous le savons. Les vols de moucherons, les sauts des sauterelles, les cohortes des fourmis' (p. 212), 'Nous savons bien ce qui se passe dans la nuit' (p. 213), and 'l'air vibre dans les toiles invisibles, vibre jusqu'à nos ventres, et nous connaissons l'histoire qui se déroule' (p. 215).

The text may be viewed as a poetic expression of Uexküll's concept of the Umwelt in *Milieu animal et milieu humain*.[18] Like the tick, spiders construct their own environment or territory and, at the same time, interact with that of other animals in a structure that Deleuze and Guattari liken to that of counterpoint.[19] Indeed, the story can be compared to a musical score and viewed as a poetic illustration of the relationship between 'le sensible et l'intelligible'.[20] For example, the spiders are also able to communicate with plants and recognise their emotions. As they inform us: 'Nous savons bien d'autres histoires, des histoires bizarres et secrètes, les pétales de fleurs qui s'ourlent, les peurs des sensitives, les barbes empoisonnées des orties' (p. 212). As Agamben has also written, the spider's web becomes a mode of communication with all forms of life.[21]

15 Deleuze writes, 'L'animal est l'être aux aguets'. See *L'Abécédaire, letter A comme Animal*, dir. Pierre-André Boutang and Michel Pamart (Sodaperga Production 1988–89). Bailly cites the example of the bat: 'Ce qui est en jeu avec lui, c'est une intensification ou une condensation de la vivacité hypersensible propre à tout champ d'action animal.' See *Le versant animal*, p. 100.

16 Henri Bergson, *L'Évolution créatrice*, 1907 (Paris: Presses Universitaires de France, 1948), p. 192.

17 Ibid., p. 273. Affinities can also be detected with the presentation of insects in Patrick Chamoiseau's *Les Neuf Consciences du Malfini* (Paris: Éditions Gallimard, 2009), pp. 154–156.

18 Uexküll, pp. 124–125.

19 See Deleuze and Guattari, *Mille Plateaux*, pp. 381–433.

20 See *Haï*, pp. 124–125

21 See Giorgio Agamben, *L'Ouvert: de l'homme et de l'animal* (Paris: Éditions Payot et Rivages, 2006), pp. 65–70.

Importantly, as spiders possess very poor vision, sound becomes not only the primary means of communication with animal and plant life but also with the cosmic, an expansion of their territory.[22] The spiders cannot see the sun, but they can hear and feel it:[23] 'Nous aimons écouter ses bruits. Le soleil craque et se froisse et fait de drôles de bruits d'insectes, vous comprenez, des bruits crissant de carapaces d'insectes, ou alors du verre qui se fendille' (p. 211).

Reference is also made in this context to the stars: 'Les Pléiades en haut, La Grande Ourse et Les Gémeaux, en bas, nous' (p. 210). Here the terms 'en haut' and 'en bas' suggest a strong spatial awareness – a thinking in relationships – confirmed by recent research into the minds of spiders.[24] Indeed, the presentation of spiders throughout the story is supported by the biologist Michael J. Rodgers, who in his *Araignées de France et d'Europe* notes that spiders have a highly developed nervous system, an excellent sense of touch and are very sensitive to vibrations and currents of air. Indeed, they rely very heavily on sound and vibration for physical communication.[25]

The Senses and the Narrative

At the same time, underpinning sensorial activity and the movements of the spiders is a fundamental intentionality or narrative logic.[26] Although animals do not possess a linear sense of time, their language conveys an

22 As Le Clézio writes in *Haï*, by imitating the sound of animals, 'l'homme ouvre un passage en lui-même. L'invisible peut venir' (p. 80). The relationship of insects to the cosmic recalls a passage in Le Clézio's *L'inconnu sur la terre*, pp. 39–40.

23 As ethologists have pointed out, spiders have very limited vision. Le Clézio's presentation of the non-human world is anchored in a biological/ethological reality. See 'How spiders see the world' <https://australian.museum/learn/animals/spiders/how-spiders-see-the-world/>.

24 See F. Punzo: 'The Capacity for Spatial Learning in Spiders: A Review', <http://www.Britishspiders.org.uk/bulletin/130301.pdf>.

25 See *Araignées de France et d'Europe* (Paris: Delachaux et Niestlé, 2014), p. 18.

26 All life forms have a sense of the future. See Kohn, *How Forests Think*, pp. 77–78.

awareness of the goal and possible consequences of their actions. The spiders refer to 'notre rôle sur la terre' (p. 215) and, as Le Clézio has earlier warned us, 'il n'y a pas de douce harmonie'.[27] They are aware of the need for continual creative expression in order to protect 'les petits animaux' against predators. The text is anchored in the structure of the quest which, for the semiotician A. J. Greimas, is that of storytelling itself with its origin in myth.[28] In semiotic terms, the spiders are presented as actantial subjects of a quest whose ultimate aim is to ensure that the cycle of life continues, that night follows day, thus ensuring the equilibrium between opposing forces, the expression of creative energy.[29] The anti-subject or 'enemy' are the forces of destruction, those of the night, 'les oiseaux de la nuit, les carnassiers de la nuit'. In other words, the text presents a conflict or struggle, the fundamental structure not only of story but of life itself.

According to Greimas, then, there are three principal stages in the unfolding of any narrative:

(a) The qualifying trials or stage of competence ('pouvoir faire', 'vouloir faire', 'savoir faire'). These are enacted in 'Nos vies d'araignées' in the sensorial categories of listening and feeling, in a constant alertness: 'nous autres, nous veillons, nous restons vigilantes' (p. 214). Without this exercise of the senses, the story cannot unfold. And, as Le Clézio informs us, silence itself can be a language.

(b) The decisive trials or performance. This is the crucial act of spinning itself, that is, the process of continual creation needed to combat the forces of destruction (the anti-subject) and to ensure that the young remain asleep and that day follows night. The

27 See *Haï*, p. 86.

28 Greimas has recognised the influence of Lévi-Strauss and of the Russian folklorist Vladimir Propp. As Greimas explained at a conference (Birkbeck College, University of London, 1985): 'I took a little Lévi-Strauss and added some Propp.' Le Clézio also refers to Propp in his essay titled *La Fête chantée et autres essais de thème amérindien*, where he comments on the impact of his narrative schema on the modern novel. See p. 174.

29 See also Chapter 1, p. 19, Note 9.

act of repetition has a narrative function, highlighting the need for constant expression: 'Nous tissons sans cesse, nous tissons la nuit' and 'Nous tissons le silence' (p. 216).

(c) The glorifying trials or outcome. The penultimate sentence suggests a successful completion of the quest: 'Bientôt le jour se lève, le soleil sort de l'eau et recommence sa course autour de la vallée' (p. 216). The story ends with a statement evoking the central role of spiders in maintaining the ecological balance necessary to sustain life on earth: 'Nous serons ici jusqu'à la fin du monde'.[30]

This narrative trajectory would necessarily imply a sense of the future, a faculty that, as we have seen, is held in common by all animals/insects. At the same time, the unfolding of the story suggests a capacity for memory and would indeed support the findings of the ethologist Martin Giurfa, in his study of spiders.[31] The text can be considered a particularly powerful rendering of the fusion of the poetic and the biological.

The Sociopolitical and the Ecological

Le Clézio's portrayal of the spider can be seen to call into question stereotypical perceptions of this animal in the West. Regarded by many as vermin or as fundamentally evil, the spider is frequently the victim of human cruelty and the target of mass extermination.[32] Indeed, as Jean-Marc Drouin explains, the assumption is made that, because spiders/insects are

30 'La cohérence entre le diurne et le nocturne exprime ce nécessaire équilibre entre les contraires, cette harmonie de la vie et de la mort, du bonheur et du mal.'
See J. M. G. Le Clézio, *Le Rêve Mexicain* (Paris: Éditions Gallimard, 1988), p. 227.

31 *Le Monde*, 11 août 2002, p. 15.

32 The German Romantics viewed the spider as essentially evil. See, for example, Jeremias Gotthelf's *Die schwarze Spinne* (The Black Spider). Elizabeth de Fontenay speaks of 'la douleur des mouches et des araignées', in *Le Silence des bêtes* (Paris: Fayard, 1998), p. 677.

so small and have minute brains, they do not suffer.[33] A number of literary devices are employed here to subvert this nature/culture opposition.

In the first instance, the spiders are accorded the right to tell their own story; they have become, as it were, empowered. The choice of the first-person plural and of the present tense heightens the effect of immediacy, drawing the reader into the text in an act of identification. Like the Indigenous peoples in *Haï*, readers shed their anthropomorphic skin in a 'becoming-animal'. Indeed, the text highlights the key role of the imagination and of the senses in closing the gap between human and non-human, conferring a central ecological role on literature and the arts.

Secondly, the narrative is not focused on conflict with other species or on power struggles, arguably the traditional representation of animals in Western literature and film. It is concerned, rather, with patterns of communication between the different forms of life, with the web of life: plants, leaves, butterflies, grasshoppers. Indeed, the spiders as receptacles of sound are engaged in a process of constant deterritorialisation: 'Nous sommes sur tous les points sensibles de la terre, et chaque onde qui parcourt les rochers, les arbres, les herbes, va jusqu'à nos pattes et entre dans nos corps. C'est ainsi que nous écoutons le monde' (p. 212). Moreover, their primary goal is to protect the young against 'les carnasssiers de la nuit' (p. 215). Indeed, far from focusing on patterns of aggression, the text brings to the fore their caring, maternal qualities, their tenderness towards 'les petits animaux'.[34]

In this concern with communication and with the need to maintain the cycle of life, the activity of the spiders acquires a central ecological significance. Like the Indigenous peoples in *Haï*, they do not take life on earth for granted: they are aware of the fragility of the equilibrium between life and death, day and night. Indeed, the night is presented as particularly dangerous: 'Vous le savez, c'est toujours un peu terrible chaque fois que le soleil disparaît et que la nuit avance sur la terre' (p. 214).[35]

33 Drouin, *Philosophie de l'insecte*, p. 179.

34 Entomologists have commented upon the strong maternal instincts and selfless behaviour of spiders. See Blanchard, *Les Araignées*, p. 42.

35 As Bailly writes, night-time for animals, 'c'est comme si l'on glissait à la surface d'un monde métamorphosé, empli de frayeurs, de mouvements effarés, d'écarts silencieux'. See *Le versant animal*, p. 10.

Furthermore, the spiders are located at the very heart of creation, at the meeting of earth and sky, sea and stars. Indeed, they are the embodiment of *le vivant*, of primordial energy, as we have seen, a core theme of Le Clézio's writing:[36]

> C'est là qu'est le centre de la terre et du ciel, même de la mer et des astres, parce que nous sommes les bêtes étoilées sur le monde. Ça c'est un peu difficile à expliquer, et d'ailleurs ça n'a pas à être expliqué. Simplement, une fois pour toutes: les Pléiades en haut, la Grande Ourse et les Gémeaux, en bas, nous. (p. 210)[37]

This relationship from earth to sky can be regarded as mystical and is central to all four texts discussed in the present book. Importantly, Le Clézio makes it clear that it is only through language, through poetic expression, that this relationship can be experienced.

In addition, in the very image of the spider's web, Le Clézio is calling into question the notion of a hierarchy of different forms of existence, positing, like Uexküll, the theme of absolute equality. Each form of life, however insignificant or small, plays a part in maintaining the equilibrium at the heart of creation.[38] Interestingly, in an interview with Isabelle Roussel-Gillet,[39] Le Clézio draws attention to the importance of insects in his work. Here, he accords them equal status with human. They are, as it were, 'fellow travellers': 'C'est un thème récurrent chez moi, les insectes nous accompagnent, sont des interlocuteurs, angoissants parfois pour certains'. He is also acutely aware of their crucial ecological function in maintaining the cycle of life: 'Un peu, sans vouloir me comparer à lui, comme Einstein qui évoque que le jour où les abeilles disparaîtront nous n'aurons qu'une

36 'C'est que le territoire ne se contente pas d'isoler et de joindre, il ouvre sur des forces cosmiques qui montent du dedans ou qui viennent du dehors, et rend sensible leur effet sur l'habitant.' See Deleuze and Guattari, *Qu'est-ce que laphilosophie?*, p. 176.

37 The relationship to the stars is a recurrent theme in several of Le Clézio's novels. See, for example, *Le Chercheur d'Or* (Paris: Éditions Gallimard, 1985) and *Ourania* (Paris: Éditions Gallimard, 2006).

38 Interestingly, in the focus here on smaller animals, Le Clézio is overturning traditional Western images of power expressed as size and strength.

39 See Isabelle Roussel-Gillet, 'En dialogue avec Jean-Marie Gustave Le Clézio', *Roman 20–50* no. 55 (2013), p. 136.

semaine à vivre'. He adds, 'Voir des insectes me rassure. Tous les insectes, les moucherons, sont utiles au fonctionnement de cette mécanique'. Indeed, this story acquires contemporary relevance in the light of the recent rapid decline in the number of insects and of the threat this poses to life on earth.[40]

The story 'Nos vies d'araignées' can be considered, then, particularly effective in closing the gap between human and non-human, in a becoming-animal. It could illustrate Deleuze's belief in literature as a revolutionary force. Indeed, it gives concrete expression to Le Clézio's statement that 'les animaux peuvent nous enseigner quelque chose'.[41] Animal thought presents itself as a constant activity, a ceaseless process of creation in a construction of relationships or networks described by Donna Haraway as 'tentacular thinking'.[42] The act of spinning here highlights the crucial role of the body and the senses in liberating these creative forces of connectivity, of community in the face of intense opposition, mirroring the political thrust that is at the core of much of Le Clézio's fiction. In her book *Eichmann in Jerusalem: A Report on the Banality of Evil*, the philosopher Hannah Arendt describes Adolf Eichmann's behaviour – his role in the gassing of Jews – as an expression of 'the banality of evil'.[43] She views him not as a monster but as the embodiment of what she terms 'thoughtlessness'. As Haraway states in her summary of Arendt's thought: 'Function matters, duty matters, but the world did not matter for Eichmann.' He lacked what spiders possess in abundance, 'response-ability'.[44] Interestingly, the teacher and writer Fernand Deligny draws upon the image of the spider to explain his work with a group of autistic children.[45] He highlights in this context

40 See, for example, the article 'Les insectes disparaissent partout sur la planète', *Le Monde*, 4 février 2019.

41 See Introduction, Note 2.

42 See Donna J. Haraway, *Staying with the Trouble: Making Kin in the Chthulucene* (Durham and London: Duke University Press, 2016), pp. 30–32.

43 Hannah Arendt, *Eichmann in Jerusalem: A Report on the Banality of Evil* (London: Penguin Classics, 2006). First published in German 1964.

44 See Haraway, p. 36.

45 See Fernand Deligny, *The Arachnean and Other Texts*, trans. Drew S. Burk and Catherine Porter (Minnesota: University of Minnesota Press, 2013). His map of 'lignes d'erre' (Maps and Wander Lines), reproduced in 'La Revue Recherches', was according to Deleuze, the origin of his theory of the rhizome. See Sandra Alvarez

the central role played by storytelling and creativity in the production of networks of meaning or the rhizomatic. Later, in a dialogue with Claire Barnet, Deleuze recognises the importance of Deligny's work and of his concept of 'lignes d'erre'.[46]

Indeed, in its focus on community and on relationality, 'Nos vies d'araignées' can be viewed as a fundamental critique of the structures of power and of domination that underpin contemporary Western and neo-colonial societies. The text would support Alain Badiou's definition of the political as a specific mode of relating to a community, one that is not confined to a particular class or institution and that is anchored in a notion of absolute equality.[47] The reader may be reminded here of Le Clézio's novel *Ourania,* set in contemporary Mexico, where he contrasts the exploitation of the land of 'La Vallée' by capitalist property developers to life in the community of Campos, a world steeped in Amerindian values and in those of Thomas More. Here, the younger members are compared to insects and birds: 'Ils sont des fourmis, c'est Jadi qui nous explique cela. Il les appelle ses fourmis, ses abeilles, ses colibris. Il dit que nous devons tous apprendre à être petits pour devenir des humains.'[48] And in his essay *L'inconnu sur la terre*, Le Clézio expresses a similar theme: 'Vous allez un peu chez les scarabées, un peu chez les mouches, un peu chez les araignées (…). Pendant quelques secondes, vous êtes celui qui est à côté.'[49]

This condemnation of the language of power and presentation of a new way of looking at the world is conveyed in a departure from conventional literary forms and from traditional rules of grammar. As Deleuze states in his critique of Noam Chomsky's linguistics: 'Une règle de grammaire est un marqueur du pouvoir, avant d'être un marqueur syntaxique' and 'Le langage n'est pas la vie, il donne des ordres à la vie.'[50]

---

de Toledo, *Pédagogie poétique de Fernand Deligny*, https://www.persee.fr/doc/comm_0588-8018_2001_num_71_1_2087.

46 Gilles Deleuze and Claire Barnet, *Dialogues*. Champs – Essais (Paris : Flammarion, 1996), p. 155.

47 See Alain Badiou avec Aude Lancelin, *Éloge de la politique* (Paris : Flammarion, 2017), p. 41.

48 Le Clézio, *Ourania,* p. 96.

49 Le Clézio, *L'inconnu sur la terre*, p. 145.

50 See Deleuze and Guattari, 'Postulats de la linguistique', in *Mille Plateaux*, p. 96.

In the first instance, the reader is struck by the highly poetic nature of the language of 'Nos vies d'araignées', an expression of 'le poème de la nature'.[51] In a sustained focus on the *le vivant*, on the act of creation, the story could be seen to differ considerably from the rest of the collection: it is virtually devoid of abstraction conveying, as we have seen, the sensorial activity that lies at the core of animal behaviour and thought. Our attention is drawn from the outset to the musicality of the text, to the rhythm of the sentences and to the repetition of patterns and words likened to the act of spinning,[52] calling to mind Deleuze and Guattari's comparison of the spider to the artist.[53] Indeed, for Le Clézio, as earlier for Uexküll, it is rhythm that lies at the very core of life. 'Nos vies d'araignées' could be compared to a dance recalling the very first story in the collection *Histoire du pied et autres fantaisies*, where the image of the foot suggests a relationship to the earth, to the very source of life.[54] This primordial musicality is conveyed through a number of stylistic devices in an attempt, as Anne Simon writes, 'd'accéder à d'autres formes de vie, à d'autres configurations du monde à d'autres stylistiques de la vie'.[55]

I now propose to look at some of the linguistic strategies that are brought into play.

51 See Bailly, *Le versant animal*, p. 118. He describes the language of the spider as 'un fragment du *poème de la nature*'.

52 The phenomenolgist Henri Maldiney has drawn attention to the importance of rhythm in animal studies. See *Regard, Parole, Espace* (Lausanne: L'Age d'Homme, 1973). The association of insects with music is a recurrent theme of Le Clézio's poetic essay *L'inconnu sur la terre*. See, for example, pp. 162–164.

53 See Deleuze and Guattari, *Mille Plateaux*, pp. 386–387.

54 The story entitled 'Histoire du pied' describes how a dancer, abandoned by her lover, attempts to throw herself off a high building, but her foot resists. We learn that she later regains the desire to live and gives birth to a child. Like the spiders, she has overcome the perils of darkness, a new day has dawned, and the cycle of life is affirmed.

55 See Anne Simon, 'Une Arche d'Études et de Bêtes', 'Zoopoétique', *Revue des Sciences Humaines* 328 4/2017, pp. 7–16. Interestingly, the musician Bernie Kraus speaks of the music of nature, of 'les paysages de sons naturels', of 'les sonorités du monde vivant'. See Jean Claude Ameisen, *Sur les Épaules de Darwin: Retrouver L'Aube* pp. 203–205 (Paris: Éditions Les Liens Qui Libèrent, France Inter, 2014)

Stylistic Devices

Repetition, which as we have seen is a distinctive feature of the language of this story, conveys not only the movement of spinning but also the relationship of the spiders to time and space as well as to other forms of life. Several strategies are adopted, the overall effect being incantatory, magical.

Anaphora

(a) The same verbs are repeated in a series of main clauses separated simply by a comma. The impression is one of an accumulation of sensorial energy: 'Nous tissons sans cesse, nous tissons la nuit' (p. 216) and 'L'air vibre dans les étoiles invisibles, vibre jusqu'à nos ventres et nous connaissons l'histoire qui se déroule' (p. 215). The repetition here of verbs in patterns of twos or threes also serves to create a rhythmic effect. At the same time, there is frequent use of syntactic parallelism, for instance, 'ni bouger, ni parler, ni penser seulement écouter, respirer, frémir' (p. 213).

(b) Nouns are also repeated in a similar use of juxtaposition and repetition: 'Nous écoutons leur passage, les ailes qui se froissent, les ailes qui battent l'air obscur, le vent des ailes [...]' (p. 215).

Tense usage

The story is related in the present tense, that is, the time of animals. As we have seen, the spiders need to be alert open to any eventuality.[56]

56 See Jean-Christophe Bailly, *Les cinq sens* (Paris : Bayard Éditions, 2014) : 'Leur vie dépend de cette forme d'attention et de toutes les formes d'attention dont ils disposent' (p. 202).

There are also very few notations of 'before' or 'after'. As the spiders explain: 'Aujourd'hui, c'est aujourd'hui' (p. 210).

Connectors: Coordinate conjunctions

(a) A similar effect is conveyed in the frequent use of the coordinate conjunction 'et', rendering the spiders' perception of time as a series of moments, each of equal value, a series of intensities:[57]

'Nous sommes sur tous les points de la terre, et chaque onde qui parcourt les rochers, les arbres, les herbes, va jusqu'à nos pattes et entre dans nos corps' (p. 212).

'Mais nous tissons des murs et des rideaux légers et nous les capturons' (p. 213).

(b) The conjunction 'puis' can have a similar serial effect, highlighting the act of movement rather than time: 'Nos routes [...], elles vont droit d'un bloc de pierre à une branche, puis à une autre branche, puis à une herbe, puis à un caillou pointu, puis à une racine, et retour au bloc de pierre' (p. 212). There is a virtual absence of connectors such as 'ensuite' or 'pourtant'.

Prepositions

The text draws our attention to the spatial position of the spiders, that is, to their relationship to the primordial forces of earth and sky. We note here the recurrent use of the prepositions 'dans' and 'entre'. The effect is one of a litany or a song. For instance, the spiders are described as 'cachées [...] entre les herbes' (p. 211); 'dans nos creux entre les herbes et les pierres' (p. 211), and 'dans nos cachettes entre les ronces, dans nos nids, dans nos crevasses' (p. 211), a description that would echo the opening of

57 Bailly writes that 'tout animal est un commencement, un point d'animation et d'intensité, une résistance'. See *Le versant animal*, p. 145.

Wajdi Mouawad's story about fireflies.[58] Moreover, two pages later: 'Nous les avons entendus là où nous sommes, dans nos creux entre les herbes et les pierres' (p. 213). There are also references to height, to above and below: 'le vent en haut des arbres'; 'en haut dans le ciel' (p. 211) and 'Les Pléiades en haut [...], en bas nous' (p. 210). The focus here on prepositions would suggest connectivity or relationality – the web of life – the central theme of the story. The text substantiates recent research into the spatial awareness of spiders, lending an entomological basis to Le Clézio's presentation and confirming the fusion of the poetic and scientific that characterises the author's rendering of the natural world.[59]

Use of analogy or synesthesia: 'Comme', 'comme si'

We may also note the use of associative thinking suggested in the expression 'comme si', conveying the relationship of the spiders principally to the cosmic. Le Clézio writes that the spiders cannot really see the sun: 'Le soleil, nous ne le voyons pas' (p. 210) and that their vision generally is very poor. Instead, he draws attention to the central role of synesthesia in making sense of the world: the spiders liken their apprehension of a weakening evening light to the act of extending their webs. In other words, light is interpreted in terms of movement: 'Le ciel est gris, c'est comme si nous étendions nos murs, nos lambeaux légers qui flottent d'une roche à l'autre' (p. 214). The sun beginning to set is experienced as both sound and movement: 'au moment où le soleil bascule de l'autre côté de l'horizon, derrière la mer et les montagnes, on entend comme un cri, ou une sorte de vague de silence blanc qui avance à toute vitesse' (p. 214). The crucial moment

58 See Wajdi Mouawad's short story 'Lampyris Noctiluca' : 'Nous sommes une multitude aux abords du chemin herbeux, blottis au creux des cailloux ou dans les feuillages.' It features in his book *Anima* (Montréal, Québec : Leméac, 2012), pp. 326–327.

59 In his study of spiders, the ethologist Martin Giurfa refers to 'les centres associatifs, où plusieurs informations sensorielles convergent pour l'action, où s'établit une forme de "dialogue" entre les informations sensorielles (vision et olfaction par exemple)'. See *Le Monde*, 11 août 2007.

when the light disappears and the spiders cease spinning is experienced as weightlessness, as emptiness: 'C'est comme si on ôtait un poids de l'air et que soudain le vide de l'espace était proche, le vide glacé du fond du ciel, là où brillent les Pléiades' (p. 214).

Use of lists: Juxtaposition

The power of thought, that is, the ability of spiders to understand and communicate with multiple forms of life, is conveyed through the use of lists, through the juxtaposition of different species of insects and of plants. The effect is again one of an accumulation of energy, that of sound and movement, an expression of *le vivant*:

> Les vols de moucherons, les sauts des sauterelles, les cohortes des fourmis, les escalades des hannetons et des crabes, les grignotements des carias, les colonies suintantes des pucerons, les danses des papillons jaunes et des libellules bleues. (p. 212)

At the same time, through the very device of listing, Le Clézio is giving poetic expression to Uexküll's notion of an Umwelt, namely that of a multiplicity of different interacting worlds, each of equal importance.

Sound of words

Like Deleuze, Le Clézio breaks down the distinction between language and sound. As Claude Cavallero comments in relation to Le Clézio: 'Par l'acte de phonation, le langage signifie autrement et sa *musique* elle-même fait sens.'[60] A number of devices are brought into play:

(a) Repetition of sounds. We note, for example, the frequent repetition of the sound 'on'. For instance: 'Nous ne dormons pas. Non, non, nous écoutons intensément, nous sentons très fort ce

60 Claude Cavallero, *Le Clézio, témoin du monde* (Paris : Éditions Calliopées, 2009), p. 137.

qui arrive' (p. 214); 'Ce n'est pas la peur, mais nous ne bougeons plus, nous ne pensons plus' (p. 215) and 'Mais c'est une odeur imaginaire, car nous ne dormons pas, nous ne dormons jamais' (p. 215). The repetition here conveys intensity, the urgency to act; life on earth cannot be taken for granted.

(b) Onomatopoeia. Here the sounds of words echo the sounds of nature. The sun is experienced as a crackling effect likened to the shells of insects or broken glass: 'Le soleil craque et se froisse et fait de drôles de bruits d'insectes, vous comprenez, des bruits crissant de carapaces d'insectes, ou alors du verre qui se fendille' (p. 211). The spiders themselves become little suns, sparks of fire, that is, pure energy: 'Et là nous sommes au centre du réseau de petites fêlures, nous devenons peut-être de petits soleils à notre tour, cachées entre les herbes, cachées et chaudes, de petites étincelles craquantes, de petites boules de feu' (p. 211). The repetition of the sound 'ie' (petites) throughout the text may convey the short, sharp chirping sounds of the other insects and animals. And the very use and repetition of the word 'vibrer' evokes sound as it resonates on the bodies of the spiders. In addition, the reiteration of short sharp sounds, especially at the beginning of words – 'les sauts des sauterelles' – may communicate the jumping movements of the grasshoppers.

Orality

As the above examples suggest, Le Clézio's language departs radically from many of the conventions of traditional literary discourse. Indeed, 'Nos vies d'araignées' would illustrate what Le Clézio has termed, 'la parole vivante du conteur'.[61]

The text is characterised by several storytelling techniques that heighten the sense of immediacy, closing the gap between text and reader. As we

61 See *Le Monde*, no. 12629, 6 septembre 1985, p. 13.

have already noted, there is very little conventional explanatory discourse. A number of rhetorical devices are brought into play:

(a) There is a direct address to the reader, a frequent use of the pronoun 'vous': 'Vous comprenez' (p. 211); 'Vous le savez, c'est toujours' (p. 214) and 'Comme vous le savez' (p. 214). This can be viewed as a persuasive device, drawing the reader into the conversation.[62] It could also be seen as a mode of bridging the gap between human and animal, of 'axing the frozen sea within us', in the words of Franz Kafka.[63]
(b) In the author's choice of using the present tense throughout, the reader is sharing the same space and time as the spiders, an effect that is reinforced by the deictics 'ici' (pp. 209, 216) and 'là' (pp. 211, 213, 216). The experience is one of pure presence.
(c) There is frequent use of the conversational marker 'c'est'. Examples include: 'C'est là que nous vivons' (p. 209); 'C'est là qu'est le centre de la terre et du ciel' (p. 210); 'c'est pourquoi nous buvons' (p. 212); and 'c'est ainsi que nous écoutons le monde' (p. 212).
(d) There are also abbreviated statements or colloquial expressions: 'Pas grand chose d'abord' (p. 214); 'Oui, il se passe beaucoup de choses' (p. 213) ; and 'Simplement, une fois pour toutes: les Pléiades en haut, La Grande Ourse et les Gémeaux, en bas, nous' (p. 210).

The choice here of an oral register mirrors the privileged role accorded the storyteller in Le Clézio's fiction.[64] Indeed, 'Nos vies d'araignées' can be

62 Referring to the reader, Le Clézio comments : 'Oui, je sollicite sa participation – comme j'imagine qu'un compositeur requiert celle des futurs auditeurs dans une composition polyphonique' See *Europe*, 'Les Marges et l'Origine', p. 31.

63 'A book must be an axe for the frozen sea within us.' See Franz Kafka, Letter, 27 January 1904.

64 An example here would be the Creole Aunt Catherine, in the novel *Révolutions*. The central role of oral storytelling in the presentation of the language of animals is explored by David Abram in his book *Becoming Animal* (New York: Vintage

read in the context of the writer's engagement with non-Western cultures and in particular with Creole communities. The story may call to mind Le Clézio's book *Sirandanes*, a collection of riddles drawn from the oral culture of Mauritius and translated by Le Clézio and his wife, Jémia.[65] In the introductory chapter where he explores the theme of animism, Le Clézio refers to 'la grand-maman araignée' (p. 15). He presents the following riddle:

> 'Ma grand-maman fait un pont, elle seule peut passer dessus?'
>
> – L'araignée (p. 34)

Indeed, as Le Clézio points out, animals are considered to be related to people: 'Ici, les animaux, comme dans les premiers contes, sont parents de l'homme, son grand-papa, ou sa grand-maman' (p. 14).

Later, the author goes on to strongly condemn the privileging of written over oral cultures:

> Ce qui me frappe [...], c'est la grande pauvreté des formes modernes. Comme si, en accordant aux seuls écrivains un pouvoir sur leur langue (parfois jusqu'à l'hermétisme), la societé contemporaine, la plus conformiste que l'homme ait jamais connue, avait délibérément limité le pouvoir de l'imagination. (p. 65)

Le Clézio returns to this point in his Nobel speech in 2008. He exhorts us to 'revenir à une communication plus immédiate, plus directe [...]. Écrire, imaginer, rêver, pour que ses mots, ses inventions et ses rêves, interviennent dans la réalité, changent les esprits et les cœurs, ouvrent un monde meilleur.'[66]

Books, 2020). He argues for a regeneration of orality as a means of bringing us closer to our senses.

65 J. M. G. and J. Le Clézio, *Sirandanes* (Paris : Éditions Seghers, 1990). Le Clézio states that these riddles continue to play an important role in the education of children in Mauritius. See p. 18.

66 See Le Clézio : 'Dans la forêt des paradoxes', 2008.

Intratextuality

It could be argued that that our reading of 'Nos vies d'araignées' is greatly enriched when viewing the text not in isolation but as part of a wider web. In an interview with Isabelle Roussel-Gillet, Le Clézio suggests a common thread behind the nine stories in the book:

> Oui, ces nouvelles s'adressent à quelqu'un mais ce qui devient intéressant c'est que je ne sais pas à qui cela s'adresse. Ce continuum a une raison d'être, c'est en inventant des histoires qu'on arrive à désamorcer des moments dramatiques ou au contraire à susciter une réaction pour que cela bouge: soit dans le sens de l'échange, soit dans une prise de conscience des injustices, des travers de la société. Ce n'est jamais détaché de la société, d'un but social.[67]

I shall now focus on the central role of insects in Le Clézio's short story 'Personne: sur une pensée de Ludwig Wittgenstein'. In a poetic exploration of the philosopher's concept of being and non-being, the text can be seen to throw further light, both philosophical and political, on Le Clézio's works and, in particular, on the theme of *le vivant*, of the birth of life and of language.

Set in Tripoli, Lebanon, during the occupation by Syrian forces (1992), the story is loosely divided into sections evoking the everyday life of couples in the city and the violent death of a pregnant female partner. The narrative moves in and out from a third-person perspective to that of a fetus struggling to be born. It is this internal monologue that will emerge as the principal focus of my study.

The opening pages recounting the death of a pregnant mother, Aliyah, introduce the central theme of the text, the fetus's longing for form of Being. This is conveyed in the recurrent motifs of desire and of potentiality: 'J'ai un désir de naître, de me former, de me fermer' (p. 312); 'Le désir d'être. C'est ce que je suis' and 'Je sais que ce qui n'existe pas encore, mais en attente, existe déjà' (p. 313). The narrator speaks of 'l'idée de cette vie' (p. 313), of a dream of life, or a creative force that precedes birth and that

67 See *Roman 20–50 : Revue d'étude du roman du XX siècle*, no. 55/2013, p. 146.

lies buried in the sky: 'Alors cette vie à venir, non pas en gestation, mais l'idée de cette vie, son rêve, son désir sont encore enfouis dans les plis du ciel' (p. 313). This potentiality is conveyed in the image 'L'Empyrée du ciel', or 'Dieu du Ciel' where everything is double (pp. 312–313).[68] However, in spite of a striving for form, the fetus or 'creature' is unable to access primordial energy, the very ground of life. In a quest to find a physical shape, it experiences a floating effect: 'La créature flotte encore invisible. La créature parcourt les possibles. Ici et là' (p. 313). This is conveyed in the sentence structure, in the use of short sharp clauses as well as in that of repetition.

In a further section, the creature experiences what seems like a call to life, like flickerings of Being. These are conveyed in the image of insects, in sensations of sound, movement and touch: 'Un bruit d'ailes. Un flottement. Un frôlement' (p. 315). The rhythm of the sentence, the use of verbal nouns and of onomatopoeia conveys the fleeting, fluttering nature of the sensorial experience. The fetus interprets this as a voice or song of life vibrating and resonating throughout its body. Its musicality is conveyed in the use of repetition:

> Le chant immédiat, l'air qui passe dans une gorge et vibre dans les cordes, dans les herbes, résonne dans une poche d'air, un crâne, une grotte, résonne dans un ventre. C'est la voix qui appartient à la vie, ou la vie qui invente la voix. Est-ce que j'existe dans le silence? (p. 315)

A few lines later, the text returns to the image of the insects: 'Un vol d'ailes légères, un frottis d'élytres, une caresse sur les nappes de l'air' (p. 315). The sensorial is now clearly associated with the advent of language, with the individual sounds of a name: 'Déjà les syllabes d'un nom, peut-être une entrée, une entrée dans la destinée [...]. La lettre <l>, la plus suave, mouillée, humide d'une respiration, avalée, déglutie' (p. 316).

Later, this desire to be born is expressed in the image of the stars, recalling 'L'Empyryée du ciel' and developing a connection with the story 'Nos vies d'araignées': 'Sous la lumière des étoiles, dans la nuit. Je voudrais naître dans la nuit. L'être boit à la source de la nuit. Boit la douceur de

68 'Illimité, indéterminé, non pas asexué, mais en formation, où tout serait double, où tout serait possible' (p. 312).

l'ombre, le lait des étoiles' (p. 316). The use of repetition here heightens the need for fusion with the cosmic, with the very source of life. However, this desire for Being is shattered: the insects, like the creature, fall victim to the bombs: 'Alors ce bruit d'ailes, ce froissement, c'est celui des papillons de nuit, [...] un tourbillon de mort quand les insectes se précipent dans les flammes' (p. 316).

Towards the end of the story, a further death is evoked in a similar configuration where the unborn is condemned to an eternal state of non-being. The passage evokes a deep sense of loss, conveyed in the repetition of 'non plus' at the beginning of sentences as well as in notations of absence, notably of beauty and of light. In stark contrast to the insects in 'Nos vies d'araignées', the fetus is unable to overcome the forces of destruction; the stars themselves have died. There can be no relationship to the cosmic, to 'L'Empyrée du ciel':

> Je suis près de la peau de l'être, je devine ce qu'il y de l'autre côté mais je ne peux franchir le seuil. C'est une porte interdite, elle ne s'ouvrira pas, elle ne s'ouvrira plus jamais. Ce qui m'entoure n'a pas de nom, c'est l'ombre d'une présence, la rumeur du non-être. Non plus un frôlement, un battement d'ailes, mais un tremblement, la peur, l'horreur, le froid, malgré le soleil, malgré le désert qui s'embrase. Non plus la nuit douce où rayonnent les astres, ces ventres pleins, tièdes, rougeoyants. Non plus l'éclat de la beauté quand un des astres se décroche de la voûte et tombe sur la terre en traçant une ligne de feu qui reste dans la mémoire. (pp. 325–326)

However, the last two paragraphs of the text introduce the unexpected: we learn that a mother and her newborn child have escaped the bombing. The baby is lying in the courtyard and laughing as a fly alights on its body (p. 329). The cycle of life has been restored.

The story can be interpreted here in the light of the final sentence of the text, a quotation from Wittgenstein's *Tractatus Logicus-Philosophicus*: 'Le monde est ce qui a lieu.'[69] Indeed, language, the sound of laughter and the movement of the fly, is associated in this paragraph with the birth of life, with an expression of the new. As Le Clézio writes: 'L'être n'a qu'une histoire, toujours la même, jamais semblable, toujours nouvellée' (p. 329).

69 Ludwig Wittgenstein, *Tractatus Logico-Philosophicus* (London and New York: Routledge & Kegan Paul, 1961), p. 5.

The baby has been released, as it were, from the tunnel of history, from 'le monde mécanique créé par les humains'[70] and from the constraints of 'what has already been said and done'.[71] Le Clézio is drawing attention here to the act of pure showing, one that Wittgenstein has described as 'the mystical element, silent and devoted to the pure act'.[72]

Interestingly, in an interview with Isabelle Roussel-Gillet shortly after the publication of *Histoire du pied et d'autres fantaisies*, Le Clézio explains why he introduced the image of the baby at the end of this story:

> C'est une des idées qui me rend optimiste: qu'on oublie, que l'on ne garde pas le souvenir des erreurs de ceux qui nous ont précédés, que chaque génération doit recommencer à partir de zéro. C'est pourquoi j'ai mis ce bébé à la fin, car je crois que c'est une nouvelle optimiste malgré le drame qu'elle décrit.[73]

In other words, each generation can shape a new world, can begin again. Le Clézio is questioning the traditional linear pattern of history with its roots in Cartesian reason and in a privileging of the referential. Language is essentially an act of creation, or series of beginnings, mirroring the cosmic circle evoked earlier in 'Nos vies d'araignées'. And, as this chapter has also illustrated, it is this language that is shared by human and non-human animals and that renders possible human–animal communication. Interestingly, in 2008 Le Clézio compared the activity of the writer to that of the spider within its web:

70 See Le Clézio's story 'A peu près apologue', where the passengers are caught up in an underground train, in a 'Vision en tunnel qui m'abstrait du réel et me place dans un état d'apesanteur – d'irréalité' (p. 335).

71 'We cannot infer the events of the future from those of the present.' See Wittgenstein, *Tractatus Logico-Philosophicus*, 5.136, p. 5.

72 'There are indeed things that cannot be put into words. They make themselves manifest. They are what is mystical.' See *Tractatus Logico-Philosophicus*, 6.522, p. 89. Alain Badiou refers in this context to 'the mystical element, silent and devoted to the pure act'. See Alain Badiou, *Wittgenstein's Antiphilosophy* (London: Verso, 2019), p. 135.

73 See Isabelle Roussel-Gillet, *Roman 20–50:* 'En dialogue avec Jean-Marie Le Clézio', p. 146.

> Il me semble que les romans tissent quelque chose, qu'ils sont une sorte de tapisserie qui se fait, se défait constamment et qu'il faut recommencer. Un peu comme le travail de l'araignée dans sa toile.[74]

Indeed, in its focus on repetition and on the renewal of life, this chapter can be viewed as a further exploration of Le Clézio's ecological vision as introduced in *Haï*.

74 "J'écris pour essayer de savoir qui je suis" - J.M.G. Le Clézio (lexpress.fr)

CHAPTER 3

Bitna, sous le ciel de Séoul

This chapter focuses on the language of birds and on the human–animal relationship in Le Clézio's novel *Bitna, sous le ciel de Séoul* (2018).[1] It also addresses the wider philosophical and sociopolitical meanings of the text. The book is composed of seven stories recounted by Bitna, a student at a Sky University in Seoul, South Korea. They are a response to a request of a young lady, Salomé, who, suffering from a degenerative disease,[2] is no longer able to leave her room:

> Je m'appelle Kim Se-Ri, mais je préfère Salomé, je ne peux plus sortir de chez moi à cause de la maladie. J'attends celui, celle quI viendra me raconter le monde. J'aime beaucoup les histoires. Ceci est une annonce sérieuse, en échange de vos histoires je vous donnerai un bon salaire. (pp. 19–20)

The text here draws our immediate attention to the central role that will be played by the storyteller and by the life-enhancing power of the imagination.[3] As in *The Decameron*, Bitna's goal will be 'faire reculer la mort': 'il me semblait que c'était mon destin de donner à Salomé le goût de la vie' (p. 43).[4] The book addresses the question of the relationship between literature and life, and between language and the body. For Salomé, a former dancer, it is an identification with the sensorial language of birds,

1 As will become clear, the stories in the book are anchored in a close biological knowledge of the language and behaviour of birds.
2 All her bodily functions will slowly collapse except the brain and the imagination. It is known as 'Syndrome douloureux egional complexe' (SDR).
3 As stated in Chapter 1, Le Clézio dedicated his Nobel speech to the Indigenous storyteller Elvira.
4 Giovanni Boccaccio (1313–1375), *The Decameron* (Penguin Classics, 1972). Written during the pre-Renaissance period, it was first published in Italian ca. 1492 and the first English edition appeared in 1620.

with poetic, creative expression that will become a healing process, a revival of her failing senses. As Le Clézio writes in *Haï*: 'Un jour, on saura peut-être qu'il n'y avait pas d'art, mais seulement de la *médecine*.'[5]

Significantly, and perhaps unusual in oral storytelling, we are not confronted with a linear narrative. The reader is immediately struck by the complexity of the narrative structure, that is, by a network of fictional stories. Furthermore, these stories are themselves interwoven with Bitna's account of her own life in Seoul including details of her meetings with Salomé. A degree of ambiguity is maintained throughout: the text is characterised by a blurring of the lines between the imagination and the real, and between story and storyteller, calling into question the dualities underpinning Western thought. Like Italo Calvino, Le Clézio is seeking to alter the status of literature and of the imagination in the West.

Indeed, as I shall show, the presentation of birds plays a central role in the dissolution of frontiers not only between human and animal, but also between thought and the senses, known and unknown, visible and invisible.

There are three stories in the book that relate to birds. These are: 'Première Histoire contée à Salomé, Avril 2016', 'Histoire des deux Dragons, à Salomé, Fin Octobre 2016' and 'La Traversée du Pont de l'Arc-En-Ciel, pour Salomé à l'Hôpital Severance, Avril 2017'.[6] I focus primarily on the first of these stories, one that explores the relationship of M. Cho to his pigeons.

'The Story of M. Cho and the Pigeons': The Mystical

As I shall show, through an identification with the sensorial language of birds, with 'les paroles de la vie', Salomé will be seen to recover a relationship with the cosmic and with the primordial forces of life. Her journey is

5 *Haï*, p. 7. Le Clézio also writes, '*L'art* : pour guérir, pour sauver les hommes' (p. 44).

6 The first relates to pigeons, the second to two mythical birds known as Dragons, the third to a jay.

one of initiation and of transformation, a 'becoming-bird'. As Le Clézio writes, 'Juste des contes pour la faire voyager' (p. 55).

As a caretaker of a block of flats in Seoul, M. Cho is raising pigeons in the hope that one day they will return to their ancestral home in North Korea.[7] The story does not unfold in a linear fashion but is interwoven with two others, encouraging a paradigmatic reading of the text. The following pattern emerges: 'Première Histoire contée à Salomé, Avril 2016', pp. 21–35, 'Suite de l'Histoire de M. Cho et ses Pigeons, Août, 2016', pp. 93–98 and 'Fin de l'Histoire de M. Cho pour Salomé, Fin Août', pp. 116–127. There are five key stages in the life of the pigeons, presented below:

1. The emergence of life: Language and the living

Significantly, the text opens with a reference to spring and to buds opening, highlighting the act of creation: 'Au printemps, quand les bourgeons commencent à sortir et que souffle le vent de l'envie des fleurs, M. Cho Han-Soo sort les cages de ses pigeons sur le toit de l'immeuble' (p. 21).[8] This movement of the cages from inside to outside will mirror the opening stages of Salomé's own trajectory of the imagination, her journey from the confinement of her room towards communication with the natural world and the cosmic.

A few lines later, the emergence of life is associated with the birth of language conveyed in the sounds and movements of the pigeons. Through the senses of smell and touch, they are able to communicate with the world just outside their cages. The text introduces the theme of relationality, that is, the interplay of different forms of life:

7 The area where M. Cho lives is near the barracks of the American army. It is also the area to which the North Korean refugees fled at the end of the Korean War in 1953. Le Clézio has said that the story of the pigeons is a true story in an interview with France Inter, March 2018.

8 As Bergson states, 'Il n'y a pas de choses, il n'y a que des actions.' He refers to 'une exigence de création'. See *Le Moment du vivant*, pp. 27–43.

> Au printemps le soleil est déjà chaud, et dans leurs cages les pigeons sont excités par le vent tiède, par les odeurs qui montent de toutes les branches des pins alentour. Ils roucoulent et se bousculent dans les cages, ils tendent le cou pour essayer de voir au-dehors, ils oublient le quadrillage du grillage cloué sur les côtés des cages. (p. 22)

Sensorial energy is evoked here not only in the choice of verbs but also in the rhythm of the text. The juxtaposition of the two phrases in the first sentence, 'par le vent tiède', 'par les odeurs' suggests a growing intensity. At the same time, through the device of onomatopoeia, attention is drawn to the very sounds of words, facilitating the revival of Salomé's senses: 'ils roucoulent et se bousculent' may evoke the cooing sound and gentle movements of the birds, whereas the repetition of hard sounds, 'cloué sur les côtés des cages', suggests a barrier or obstacle. Salomé as listener can share the excitement of the pigeons: she is now impatient to leave her own cage, the cage of her room.

Significantly, the very act of the emergence of life is conveyed in the narrative structure and in the use of spacing. The text moves in and out from descriptions in the present tense of the birds on the rooftop to passages relating to their historical past or to the life of M. Cho as a caretaker, the transition marked only by the use of spacing. Indeed, there is very little textual continuity in the story of the pigeons. Their emergence on the rooftop unfolds as a series of beginnings mirroring the rhythm of creation, the creation of life and of language: 'Au printemps' (pp. 25, 27), 'Donc, M. Cho, ce matin de printemps' (p. 26), 'C'est le moment' (p. 27), 'Le moment est arrivé' (p. 28), and 'C'est l'heure' (p. 29). Language becomes an event, a calling into existence. As Jean-Luc Nancy has written: 'Presence is born, and does not cease being born.'[9] In other words, there is a constant regeneration of meaning, for Salomé a renewal of the senses.

Indeed, in this use of spatialisation, the text can be compared to a musical score: each time the birds appear on the roof, the reader/listener is provided with additional information, with variations on a theme. At the same time, this device functions like a breath of air; we are no longer imprisoned within 'le carcan de la représentation', within 'what has already

9 See *The Birth to Presence*, trans. Brian Holmes and others (Stanford: Stanford University Press, 1993), p. 9.

been said and done', in the words of Wittgenstein.[10] This essentially poetic device will become a dominant feature of the text, symbolised in the act of flying itself.

2. *Naming*

Furthermore, on their second appearance the following spring morning, we learn that the birds have been given names:

> C'est le matin, il n'y a pas de meilleur moment pour les pigeons. M. Cho a monté les cinq cages l'une après l'autre, dans chaque cage il y a deux couples de pigeons séparés par une cloison de carton fort. Chaque couple a un nom, un nom de famille en quelque sorte, et chaque membre de famille un prénom. (p. 25)

The linguistic register here would seem very different from that of the previous passage. The initial focus appears to be on the factual and on the uniquely visual: we note the symmetry of the second and third sentences and the repetition of the same syntactical structures. At the same time, in the very act of naming Le Clézio is exploding the human–animal dichotomy at the core of Western thought. Like the Indigenous peoples in *Haï*, M. Cho is according animals the status of people, one of absolute equality with the human. The reader may be reminded here of similar findings of Descola and Latour discussed in Chapter 1.[11] Furthermore, birds, like humans, can recognise their names. Here, as elsewhere in the book, Le Clézio is drawing upon his own extensive biological knowledge of birds.[12]

Indeed, our attention is drawn here to the central role played by the language of sound as a mode of communication between human and animal. The text may call to mind the biblical association between sound, the

10 See *Tractatus Logico-Philosophicus*, 5.136, p. 47.

11 See Descola (pp. 7, 10) and Latour (pp. 11, 17).

12 For the central role of sound in the life of pigeons, see Jennifer Ackerman, *The Genius of Birds* (London: Corsair, 2017), pp. 264–246.

emergence of language and the Creation of the world: birds were allotted names in the Garden of Eden and humans understood their language.[13]

It is then this theme of human–animal communication that will emerge as central to Salomé's own journey towards the Outside. A third evocation of the birds conveys their impatience to leave their cages:

> C'est le moment. Les pigeons l'attendent avec de plus en plus d'impatience, ils tournent sur eux-mêmes dans l'étroit compartiment des cages, ils s'essaient à battre des ailes, et le bruit de leurs rémiges émet un sifflement qui exagère encore leur impatience. M. Cho ressent cela dans son propre corps, comme un fluide électrique qui parcourt ses membres, s'exaspère au bout de ses doigts, hérisse les petits poils sur le dos de sa main. (p. 27)

Their emotions, their eagerness to fly, are expressed here in the birds' movements as well as in their sounds conveyed in the use of assonance, and rendered more intense in the accumulation of main clauses. At the same time, the rhythm of the last sentence, together with the juxtaposition of two main clauses, communicates the sensation of an electric current flowing through M. Cho's body, a feeling that will be shared by Salomé: the langage of the birds in this passage will be experienced by Salomé as pure energy, as 'un élan vital'.

Indeed, the very choice here of the image of an electric fluid would suggest a direct physical communication with the living core of matter, with the material forces of life or space of origins.[14] It is as if communication with birds can open up hitherto unexplored dimensions of the self.

It is this sensory experience that prompts M. Cho to establish a closer relationship and open a direct dialogue with the birds. He now kneels before the cages and addresses them individually by their names. This act is conveyed concretely in the spatialisation of the text: the birds are presented as a vertical list of couples, bringing to the fore the rhythm and sounds of

13 *The Holy Bible*, Revised Standard Version, Genesis 1:20–23, 26, 29.

14 All matter is alive. It is not a fixed essence but an activity. As Jennifer Ackerman explains, the direction of the flights of pigeons is determined by electric currents, by magnetic fields. See *The Genius of Birds*, pp. 260–264.

language, the affective immediacy of individual words.[15] It is both an expression of how birds think and a stimulus to Salomé's senses:

> renarde, et toi le garcon, pinson
> bleu, et toi rouge-gorge
> fusée, flèche blanche
> lumière, lune
> la mouche, la cigale
> voyageuse, président
> acrobate, petit-gris
> diamant, dragon noir,
> danseuse, sabre

The response of the birds highlights the magical power of naming: 'l'oiseau nommé cesse de s'ébattre, renverse sa tête en arrière et regarde de son œil jaune' (p. 28). At the same time, the gaze here is an address, an act of mutual recognition.[16] It conveys a message, a promise that they will fulfil his mission.[17] We may be reminded of Bailly's description of a bird, of 'Ce regard qui au lieu de nous clouer au sol suppose l'existence d'un espace de délibération et de transfert.'[18] Indeed, the look triggers in M. Cho a Proustian process of involuntary memory, a reawakening of his dream of a return to his family roots in the countryside near the border with North Korea: 'Une promesse de quoi? Il ne pourrait pas le dire, mais c'est comme ça: quelque chose qui s'unit à lui, et lui donne la mémoire du passé, quelque chose comme un rêve qui se reprend après des jours de sommeil' (p. 28). For Le Clézio, as for St Augustine, memory is a material

15 In her role as language teacher, Bitna has expressed a dislike for conventional sentence structure and grammar : 'Tandis que les étudiants planchaient sur les problèmes de syntaxe, je me laissais aller à la douce rêverie sur les mots, ce que j'ai toujours bien aimé faire', p. 134.

16 For Jacques Derrida, the whole history of Western philosophy has refused to acknowledge the address of the animal gaze. See 'Contemporary French & Francophone Studies *Human–Animal* Part 2', p. 696. See also Elizabeth de Fontenay, *Quand un animal te regarde* (Paris: Gallimard, 2006).

17 This would suggest a generational memory in birds. See Ackerman, *The Genius of Birds*, pp. 260–261.

18 See *Le parti pris des animaux*, p. 18.

substance, an opening of vast vistas of infinite time and space. As Le Clézio writes in *Révolutions*:

> La mémoire n'est pas une abstraction, pensait Jean. C'est une substance, une sorte de longue fibre qui s'enroule autour du réel et l'attache aux images lointaines, allonge ses vibrations, transmet son courant jusqu'aux ramifications nerveuses du corps.[19]

3. The opening of the cages

After fastening messages to their feet, M. Cho opens the cage of Dragon noir as 'le moment est venu de faire ce qu'il sait faire: voler' (p. 29). In her description of M. Cho's attempts to assuage the fear of the birds at leaving their cages, Bitna could be said to be preparing Salomé for the next stage in her journey of the imagination, that is, the flight of the birds over the city. M. Cho now reassures Dragon noir by choosing 'des mots doux, des mots ronds, des mots légers' (p. 30), the adjectives here evoking lightness, the freedom of flying, an illustration of the material power of individual words.[20] In addition, we learn that M. Cho 'souffle sur sa tête, sur le bout de son bec', the very act of blowing evoking the open air, whose sensorial impact is strengthened in the use of onomatopoeia.

The reaction of Dragon noir, the pebbly sounds he makes in his throat, 'les mots de son langage, du langage de sa gorge seulement', indicate that he is indeed ready to fly. Again, the minute details presented here display a close biological knowledge of birds, thereby heightening the 'effet de réel'. We are never in the world of pure fantasy. Salomé, in an act of identification with Dragon noir, is now ready to leave her own cage and accompany the birds in their flight:

> Dragon noir écoute, dans son œil jaune la pupille s'arrondit, et au fond de sa gorge M. Cho entend les petits cailloux qui roulent, ce sont les mots de son langage, du

19 *Révolutions*, p. 115. As Le Clézio also writes : 'La mémoire de la tante Catherine était sans fond [...]. C'était avant toutes les guerres où le monde était encore innocent', p. 23.

20 'Comme tout ce qu'on voit, tout ce qu'on entend, les mots sont réels.' See Le Clézio, *L'inconnu sur la terre*, p. 110.

> langage de sa gorge seulement car tout le corps de l'oiseau a envie de parler avec ses rémiges, avec ses ailes, avec les plumes de sa queue, fendre l'air et parler en plongeant dans les courants d'air. (p. 30)

The strength of desire for physical expression is rendered here in the accumulation of prepositional clauses and in multiple references to the body. Sensorial intensity is also communicated in the use of synesthesia[21] and again in that of onomatopoeia, a central feature of the language of birds and a mode of relating to the world that is particularly strong in animals: here, the very sounds of the words, 'les petits cailloux qui roulent', suggest the hard guttural sounds in the throat of the birds.

In an identification with the birds, Salomé feels that she too is ready to 'échapper à l'immobilité de son appartement, pour respirer' (p. 180). She is moving towards the space of the Open, the space of freedom. Le Clézio will now present two versions of the flight of the birds over the city: version 4a. and version 4b.

4. *The flight of the birds over the city*

The first description of this flight is preceded by a pause in the oral narrative conveying discontinuity, the sensation of leaving the ground, of entering another mode of being or relationship to the world. Indeed, we are told that, as M. Cho is waiting for the birds to return, he closes his eyes and 'il essaie d'imaginer ce que voient Dragon noir et sa compagne Diamant, au-dessus de la ville' (p. 31), a gesture that will be repeated by Salomé. This very act reminds us of the importance of the imagination: without the ability to identify with others, it is impossible to understand the non-human world. To quote Le Clézio: 'C'est le paradis des écrivains, l'imagination pure.'[22] Interestingly, there are strong echoes in the text of Calvino's story 'Palomar on the Terrace', where Mr Palomar

21 As Lévi-Strauss has written, synesthesia is an attribute of all forms of life. See Chapter 2, Note 11 where reference is made *La Pensée sauvage* and *Mythologiques*.

22 J. M. G. Le Clézio, *Ailleurs : Entretiens avec Jean-Louis Ezine* (Paris : Arléa, 1995), p. 28.

'tries to conceive the world as it is seen by birds' – and the birds are mostly pigeons.[23] And, listening to the blackbirds in another story, Mr Palomar wonders: 'But what if language were really the goal towards which everything in existence tends? Or what if everything that exists were language, and has been since the beginning of time?'[24]

4.a.

> Les hauts immeubles en verre, debout pareils à des falaises de cristal, les rubans des autoroutes, et ensuite le grand fleuve. L'énergie accumulée dans leurs ailes pendant des semaines de confinement se transforme en force électrique, les ailes battent à toute vitesse, les courants de vent les poussent vers le haut, puis les trous glacés au-dessus de la rivière les font plonger. Dragon noir guide en tête jusqu'à la rivière, puis c'est Diamant qui prend les devants, elle longe la rive jusqu'au pont, vers l'île. Il y a d'autres oiseaux dans le ciel, des mouettes, des goélands, et près de l'île, des groupes de canards. Les pigeons ne s'arrêtent pas, ils tracent des cercles au-dessus de l'eau, la surface miroite avec des frissons, les touffes d'herbe et les joncs s'inclinent dans le vent, sur le grand port les autos sont arrêtées par l'embouteillage du matin, une rumeur de klaxons, ou bien les cris des canards ou encore l'appel du train qui franchit lentement la rivière. (pp. 31–32)[25]

The focus here is on energy, the energy of movement, a sensorial quality that Salomé, confined to her room and ill, would be lacking. This is conveyed in the second sentence in the image of an electric force associated earlier with the pigeons as well as in references to speed (lines 3–4). The impression is that of a race: 'Dragon Noir guide en tête [...] puis c'est Diamant qui prend les devants'. The text would bring to mind Bailly's

23 See Italo Calvino, *Mr. Palomar* (London: Pan Books, 1986), p. 49. Both these poetic writers focus on the language of birds and bring together literature and the philosophical.

24 Ibid., p. 24.

25 The imagery would call to mind the flights of birds in Le Clézio's poetic essay *L'inconnu sur la terre*, pp. 203–204. Here he presents flight as a mystical experience, as the true source of knowledge : 'Tout est possible alors ; l'union magique de la mer, de la terre et du ciel ; le seul savoir : voler. Être dans le sein de l'air, porté, lié, balancé dans le vent, confondu avec les mouvements du ciel, comme des nuages, comme des gouttes de pluie' (p. 205). He adds : 'Je voudrais apprendre à voler haut, puis à tomber comme un silex ' (pp. 207–208).

description of flight as 'une intensification ou une condensation de la vivacité hypersensible propre à tout champ d'action animal'.[26] In addition, as the birds fly, they are subject to the power of the winds, to the cosmic forces, the very source of life: 'Les courants du vent les poussent en haut, puis les trous glacés au-dessus de la rivière les font plonger' (p. 32). As Deleuze and Guattari write: 'C'est que le territoire ne se contente pas d'isoler et de joindre, il ouvre sur des forces cosmiques qui montent du dedans ou qui viennent du dehors, et rend sensible leur effet sur l'habitant.'[27] Movement is also evoked in the image of circling, 'ils tracent des cercles', a dance movement that will characterise later flights over the mountains (pp. 97, 124) and that would suggest the cosmic circle. At the same time, the birds communicate with other forms of life, with the ducks and the plants, contributing to a sensation of *le vivant*. The overall impression is that of a musical score, an interweaving of different territories, of different themes. Like the insects discussed in Chapter 2, the birds can be compared to artists.

In addition, primordial energy is conveyed in the rhythm of the text, that is, in the accumulation of main clauses separated only by a comma as in the second and last sentences. There are no conjunctions, virtually no adjectives. It is also communicated in a repetition of the same sounds, mainly at the end of words, for instance, 'Diamant', 'les pigeons', 'les courants' and 'des goélands'.

Bitna's account has an intense physical impact on Salomé, provoking a change in level of consciousness, a sensation of freedom or out-of-body experience: 'Votre histoire, je l'aime bien, je sens que je peux voler moi aussi comme les pigeons de M. Cho, au-dessus de la ville. Je me sens si légère!' (p. 34). However, she is not totally satisfied: she would like the flight to be anchored in a specific geographical reality, in a real sense of place: 'Les noms des endroits, là où ils volent, vos pigeons, dites-moi les noms!' (p. 34). As earlier, in the case of pigeons, the focus on naming may facilitate the process of imaginative identification, enhancing 'l'effet de réel'. It may also reflect Le Clézio's desire to bring together the poetic and the concrete everyday,

26 See *Le versant animal*, p. 101.

27 See *Qu'est-ce que la philosophie*, p. 176.

the imagination and the real, as in *Haï*. Responding to Salomé's request, the reader/listener will now be confronted with a surfeit of names, some of which refer to actual places in Séoul.[28]

4.b.

> Dragon noir et Diamant ont volé au-dessus des immeubles jusqu'à la rivière Han, puis ils sont passés au-dessus de Yeouido, les grandes maisons blanches du gouvernement, les parcs où les petits vieux promènent leurs petits-enfants le dimanche après-midi, ils ont basculé sur le côté et les voici qui passent au-dessus du grand pont Seogangdaegyo, avec les millions de voitures qui courent les unes derrière les autres comme des insectes. Ils ne s'arrêtent pas là, ils passent au-dessus de l'île des canards, puis ils reviennent en arrière, ils longent la rivière, puis le canal, ils vont à Myeong-dong, au-dessus de l'hôtel Savoy, il y a plein de rues encore sombres, ils passent le long de la grande montagne, peut-être que Diamant voudrait s'arrêter un instant dans les pins sur la montagne, elle aime tant l'odeur des aiguilles, elle voudrait que Dragon noir se décide à construire un nid, peut-être, un jour, mais lui bat vite des ailes, il trace une longue courbe vers Jongno, vers la tour de la librairie Kyobo, ensemble ils volent vers Insadong, puis vers les jardins de Changgyeonggang, au-dessus du jardin secret, l'eau des petits lacs étincelle au soleil, il y a l'odeur des arbres, des fleurs, le vent qui descend de la montagne les repousse en arrière, ils sont au-dessus de Dongdaemun, de Samcheong, et M. Cho, sur le toit poussiéreux de son immeuble, peut imaginer ce qu'ils voient, les toits traditionnels en tuiles vernissées qui brillent, les jardins, les cours carrées, ensuite les pigeons reviennent près du palais de Gyeongbokgung, jusqu'à la gare, ils redescendent vers le soleil, c'est déjà la fin du jour, ils sont fatigués d'avoir tant volé, ils tracent encore un demi-cercle autour des immeubles de Samsung, et le vent de la rivière, ou bien le vent solaire les repousse vers la haute silhouette adossée à la colline du Dragon, vers le toit plat où M. Cho les attend. (pp. 34–36)

In the first instance, the very length of this passage conveys both a greater sense of space and the sensation of flight itself. Uninterrupted movement and spatial expanse are communicated in the rhythm of the text, in the second seventeen-line sentence and in the accumulation of main clauses all linked by 'puis'. The 'effet de réel' is further strengthened in

28 Hyeli Kim has written that Yeouido, Myeong-dong and Gyeongbokgung are actual place names in Séoul. See 'Les Paysages coréens dans *Bitna, sous le ciel de Séoul* comme une source de l'inspiration poétique', *Les Cahiers J.-M.G. Le Clézio*, no. 13/2020, pp. 75–86.

the numerous references to place names, to the landscape over which the birds fly and to their sense of direction: 'vers', 'au-dessus'.

At the same time, the impression given is that of a wider community: the Umwelt of the birds intersects with several other worlds, with the territories of the ducks in the river and of the trees and flowers in the mountains. Indeed, we are presented here with a much richer sensorial experience, a heightened stimulus to Salomé's failing senses, to her loss of energy. The multiple and lengthy place names with similar endings ('ong', gang, o'o, 'gung') convey intensity of sound as does the repetition of the same individual words: 'l'odeur des aiguilles', 'l'odeur des arbres', 'l'odeur des fleurs'. Indeed, the passage would give poetic expression to Merleau-Ponty's statement 'Umwelt, c'est une mélodie qui se chante elle-même.'[29]

However, it is clearly not only a question of movement and sound. In the first instance, a further sensorial enrichment is communicated in the introduction of the sense of smell, one that will play a critical role in the later flights and that indeed is very acute in pigeons.[30] Secondly, the sense of touch is also brought into play in more frequent references to the cosmic forces, in particular to the wind and its power over the birds: 'Le vent solaire qui descend de la montagne les repousse vers la haute silhouette adossée à la colline du Dragon'.

This second description of the flight produces in Salomé a state of ecstasy or trance: the very sound of Bitna's words, of the individual names, generates a more intense mystical fusion with the cosmos: she now undergoes a complete metamorphosis, an experience of deterritorialisation: she has indeed become a bird:

> Salomé avait le visage enfiévré, quand je disais les noms, elle fermait les yeux et elle glissait dans l'air avec le couple de pigeons, elle filait d'une rue à l'autre, elle sentait le courant d'air de la rivière, elle entendait le bruit mélangé des voitures, des camions,

29 See Merleau-Ponty, *La Nature*, p. 228.

30 Le Clézio is again bringing together science and the imagination. For the importance of the sense of smell in pigeons, see Ackerman, *The Genius of Birds*, pp. 266–273.

> des autobus, et aussi le fracas de métal du train qui glisse dans sa rainure près de la gare de Sinchon. (p. 36)[31]

Here the rhythm of the sentence, the accumulation of main clauses, conveys the act of flying itself: the experience is once again a revival of her senses, a process that Salomé will need to constantly repeat. Furthermore, the very closure of the eyes[32] would confirm a central theme of the text, that is, the key role played by the imagination in a journey of the self towards the realm of the non-human and of the invisible. It would also substantiate Le Clézio's earlier statement that by identifying with animals, 'l'homme ouvre un passage en lui-même. L'invisible peut venir' (*Haï*, p. 80). At the same time, throughout the story, Le Clézio is drawing upon a detailed biological knowledge of the language of pigeons including their faculty of memory and sense of the future.[33] This would facilitate the process of imaginative identification, of 'becoming-bird'.

'The Story of M. Cho and the Pigeons' is resumed fifty-seven pages later: indeed, as will become apparent, it is interwoven with several other stories, as well as an account of the personal relationship between Bitna and Salomé.

5. *The final stages of the flight*

Suite de l'histoire de M. Cho et de ses Pigeons, août, 2016.

5.a. The flight over the river and mountains

We learn that prior to the final flight over the river to his birthplace, M. Cho attaches messages to the feet of the birds:

31 In his poetic essay *L'inconnu sur la terre*, Le Clézio writes : 'J'imagine que je vole : c'est le seul rêve que je voudrais vraiment réaliser' (p. 201) and 'Le seul savoir, voler' (p. 205).

32 See also pp. 97, 123, 125.

33 See Ackerman, *The Genius of Birds*, pp. 5–17.

> mer
> *île*
> vent
> aile
> retour
> infini
> longtemps
> *caresse* (p. 94).

This technique of spatialisation recalls Le Clézio's series of articles titled *Vers les icebergs* where he speaks of words as a dance, as flight and as movement (p. 62).[34] It is a poetic device that enables Salomé to feel each individual word as the expression of energy, of the living. This initial act of preparation is followed by a description of several flights over the forest and mountains[35] towards the North, the effect being that of a ritornello. As Salomé's health is deteriorating, there is a need for constant repetition in a slowing down of the process of death and in an easing of the pain of dying.[36] For instance, we learn at this point of a training flight over the mountain highlighting the faculty of smell:[37]

> L'odeur délicieuse de l'air, les bois de pins au soleil, les petites grasses entre les pierres, et peut-être même l'odeur de la neige, une odeur calme que seuls les oiseaux peuvent sentir. (p. 97)

At the same time, attention is drawn to the energy of movement, to speed itself:

> Les deux oiseaux se retrouvèrent dans le ciel, et ils se mirent à tournoyer l'un autour de l'autre, si proches et si rapides que M. Cho craignit un moment qu'ils se fracassent sur les rochers. (p. 97)

34 J. M. G. Le Clézio, *Vers les icebergs* (Paris : Fata Morgana, 1978), p. 62.

35 The image of the mountains is associated with fusion with the cosmic, with the sacred. See Chapter 1, p. 8, where reference is made to the author's short story *La Montagne du Dieu vivant.*

36 'Elle est gravement malade, elle ne peut plus bouger de son fauteuil' (p. 99).

37 The focus on smell here is a further illustration of Le Clézio's own knowledge of the lives of pigeons, a blurring of lines between science and fiction.

Interestingly, M. Cho closes his eyes at this point 'pour mieux ressentir ce qu'ils ressentaient, comme un ouragan de lumière et de vent, faisant tourner la montagne au-dessous d'eux, tressant les fils de nuages blancs et gris' (p. 97). Hypnotised by Bitna's words, Salomé 'ferme les yeux, elle aussi', a further illustration of the transformative power of language and of the imagination.

In the final chapter of the story, 'Fin de l'Histoire de M.Cho pour Salomé, Fin Août 2016', 'evoking the flight of the birds de l'autre côté de la frontière' (p. 121), there is an increased focus on the primordial forces of wind and light (p. 123). Salomé is now undergoing a series of mystical experiences that are increasing with intensity as the birds approach their destination. Indeed, in the presentation of these final flights, the border between fiction and reality completely collapses. Under the magical hypnotic power of Bitna's words, Salomé undergoes a complete metamorphosis: she now sheds her human skin and becomes a bird flying towards 'l'autre rive du monde,' towards 'le pays du retour' (p. 123). The text would suggest the power of the human voice to communicate with the natural world, calling to mind Le Clézio's essay *Haï*. Indeed, the very repetition of words by Bitna anticipates the shamanic ceremonies associated with South Korea evoked in a later story in the book.[38]

> Écoute, le vent vient de la mer, le vent clair du matin, respire et sens le vent sur la peau de ton visage, Salomé, tu voles très haut dans le ciel, vers le nord, vers l'autre rive du monde, c'est ton dernier voyage, avec Dragon noir, avec Diamant, avec les autres, le vent t'enivre, le vent éblouit tes yeux et coupe ton souffle, mais tu continues à voler, tu vas droit vers la fin de ton voyage, tu écartes les bras et tu sens le vent sur ton corps, tu ne pèses plus rien, tu es une plume dans le vent, une feuille, un pétale de fleur, et en dessous de toi le fleuve avec ses îles te pousse vers le haut, vers le nord, vers le pays du retour. (p. 123)

As Bitna continues speaking, Salomé's senses are further awakened. The very length of the above sentence evokes not only distance but also the

38 See the story of Naomi (pp. 172–181). Mention is also made of the story of the two Dragons: the 'Dragon du nord' and the 'Dragon du sud' (p. 173).

energy of the wind that is driving Salomé on. And, as earlier, Le Clézio is drawing attention to the magical power of individual words:[39]

> Elle ouvre ses mains, elle sent l'air entre ses doigts, elle respire le vent, elle goûte le sel de la mer et le miel des praires fleuries, les longues tiges des eulalies qui ondulent dans le vent, le feuillage des arbres, les haies brillantes des camélias, et tous les chemins qui se croisent, non pas des routes, mais des sentiers bordés de murs de pierres, et les toits de tôle bleue des villages, ce sont les mots qui la transportent, elle n'a même plus besoin de les entendre, ils naissent dans son esprit comme des fusées qui s'éclairent. (p. 124)

Indeed, there is an increased focus here on images of the air and the wind as well as on the act of breathing, an activity that for Salomé is becoming increasingly difficult. And, in striking contrast to her initial sense of imprisonment within her room, flight now suggests the space of the Open, that of freedom, 'd'une pure ivresse du vivre'.[40]

5.b. The arrival at the birthplace of M. Cho, a farm on the other side of the river

In this concluding section of the story, we learn of the successful arrival of the pigeons in spite of the sporadic shooting at the border with North Korea.[41] They have indeed reached 'the other side',[42] suggesting the myth of the eternal return, the space of origins:

39 There are again echoes of Le Clézio's earlier essay *L'inconnu sur la terre* : 'Mais c'est cela que veulent dire les mots : liberté. Libre, on est libre, sans murs, sans vitres, sans barrières' (p. 109) and 'Comme tout ce que l'on voit, comme tout ce que l'on entend, les mots sont réels' (p. 110).

40 See Bailly, *Le versant animal*, p. 89. Here he describes the flight of starlings as 'une condensation de ce qui est non seulement libre mais véritablement libéré et agi dans le ciel, signature d'une pure ivresse du vivre en un battement singulier et rêveur'.

41 This refers to the hostilities between North and South Korea that persisted after the war was technically over. The relationship between the two countries has recently deteriorated. See *Le Monde*, 17 juin 2020, p. 12.

42 'The other side' may also connote North Korea.

> Les pigeons sont arrivés là, guidés par un ordre mystérieux et évident, ils parlent de l'autre monde, de l'autre côté de l'estuaire de la rivière, un monde qui cesse d'être étranger [...] C'est la fin de leur voyage. (p. 126)

Salomé's reaction to the story would point to a philosophical acceptance of her fate:

> Salomé est renversée dans son fauteuil, la tête un peu inclinée, ses yeux sont pleins de larmes, mais elle ne sait pas si c'est de joie ou de détresse. C'est la fin d'une histoire, la fin d'un voyage. (p. 126)

At the same time, however, 'The Story of M. Cho and the Pigeons' is greatly enriched when read within a vast tapestry of meanings and of interlocking stories, a jigsaw puzzle that can never be completed.

The Mystical Dimension in *Bitna, sous le ciel de Séoul*

In a recent interview, Le Clézio describes the mystical as 'une perception immédiate', as a sensuory experience of fusion with the cosmic: 'C'est le lien que l'on peut faire entre la terre et le ciel, tout ce qui relève de la relation de la terre au ciel.'[43] As we have seen, this relationship is illustrated in the flights of the pigeons over the city and mountains and in Salomé's own act of identification with the birds. At the same time, the image of the cosmic circle, of the eternal return, is depicted in the movement of the global narrative from the emergence of life (p. 21) back to its material roots or the space of origins (p. 126).

This coexistence of opposites, of life and death, is also mirrored in the choice of the couple of birds, male and female, Dragon Noir and Diamant, and expressed in the dance movements as they fly and in the very act of circling. The Dragon is in itself a mythic representation of the dualities yin

43 Justyna Gambert, 'Entretien avec J. M. G. Le Clézio', *Les Cahiers J.-M.G. Le Clézio,* no. 8, p. 67.

and yang underpinning several world philosophies: Taoism, Zen Buddhism and shamanism.

Indeed, the image of the Dragon features in the second story relating to birds, 'Histoire Des Deux Dragons à Salomé, Fin Octobre 2016'. Here we learn of a young girl, Naomi, who, like Salomé,[44] can experience the invisible: she can feel the presence in the sky of two Dragons that also take the form of 'deux serpents enroulés', recalling the image of the bird-serpent Ouroboros in Amerindian myth. Ouroboros is also the symbol of the cosmic circle.

In both stories, it is the transformative magical power of the language of Bitna, her chant-like use of repetition that lies at the core of Salomé's mystical identification with the flights and with the cosmic forces: the Dragons here are described as 'pareils à des nuages, ou pareils à un reflet d'eau sur la mer, ou encore pareils aux goutes de la pluie que tu entends mais que tu ne peux pas voir' (p. 173) and later, 'elle sent juste leur présence, c'est pareil au souffle du vent chaud en été, ou bien aux tourbillons qui emportent les feuilles d'or des gingkgos' (p. 183).

The Moral Dimension in *Bitna, sous le ciel de Séoul*

Importantly, the book is not only concerned with cosmic consciousness or the mystical. In its focus on everyday life in the city and on human relations, it possesses a further affinity with Zen Buddhism, a practice that is also widespread in South Korea. Indeed, for Buddhists, Being and action cannot be separated. And for Bitna, the storyteller, it is a question of what Gaston Bachelard terms 'l'imagination morale'.[45] As Philippe Pierron also

44 Salomé was born in 'l'année du Dragon' (p. 179).

45 'Nous croyons pour notre part, que les lignes imaginaires sont les vraies lignes de vie, celles que se brisent les plus difficilement. Imagination et volonté sont deux aspects d'une même force profonde.' See Gaston Bachelard, *L'Air et Les Songes. Essai sur l'imagination du mouvement*, 1943 (Paris : José Corti, 1994), p. 144.

writes in reference to Le Clézio, 'La Poétique a une portée éthique, sociale et culturelle.'[46]

The Buddhist values of love, care and compassion for all living things, together with a critique of all forms of violence, are exemplified in M. Cho's relationship to his pigeons and in the messages of peace and love that the birds bring to a family separated by war, by the conflict between North and South Korea.

At the same time, these values are illustrated in the personal relationship between Bitna and Salomé, the victim of egotism and of society's indifference to suffering. The text draws attention here to the ethical role played by the imagination. Indeed, at several intervals, the narrative presents a tea-drinking ritual between Bitna and Salomé: it is this ritual that fires Salomé's imagination and strengthens the relationship between the two. And it will be Bitna's stories that alleviate Salomé's pain of dying.[47] The text would call to mind a similar ceremony conducted between the Creole-speaking storyteller Aunt Catherine, and her nephew, Jean Marro, in Le Clézio's novel *Révolutions*. Here the stories of birds at Catherine's home, Rozilis, have a similar curative impact on Jean, who is suffering from an acute sense of alienation in the contemporary city of Nice.

Indeed, it is this human–animal identification that underpins much of the book and that is beautifully illustrated in the final story that Bitna tells, 'La Traversée du Pont de L'Arc-en-Ciel, pour Salomé, a L'Hôpital Severance, Avril, 2017,' (pp. 201–205). Here a girl, Naomi, is described as comforting and easing the pain of a dying bird, a young jay who has fallen from a tree and become separated from her mother.[48] The bird, named O'Jay, is distinguished by 'ses merveilleuses ailes bleues' (p. 204) and Naomi will later comfort the bird's dying days by playing recorded birdsongs, by

46 Pierron, *Les Puissances de l'Imagination*, p. 94.

47 The relationship between the two is ambiguous, a mutual dependency. As Bitna explains, Salomé was the only person who counted in her life. There is a blurring of the frontiers between literature and life, between the imagination and the everyday.

48 The canvas of intertwined relationships is broadened. Naomi herself was abandoned by her mother and adopted by Hana, who is now living with Naomi and looking after the bird.

evoking 'le goût du vent dans le ciel, [...] les montagnes et les forêts, l'odeur des pins au soleil' (p. 208).

The warmth and love displayed here by Naomi for the bird would be communicated to Salomé who, now in hospital and barely able to breathe, is rapidly approaching her own death. Importantly, as Bitna informs us, in choosing this story, she is presenting Salomé with a broader philosophical perspective, calling to mind Boethius's book *The Consolation of Philosophy*[49] or Michel de Montaigne's *Essais*.[50] As she states: 'J'ai eu envie de partager cet oiseau avec Salomé [...] parce que son histoire est commune à tous les vivants. C'est l'histoire la plus mystérieuse de la vie, avec l'instant de la naissance' (p. 206). Salomé herself accepts the inevitability of dying and of suffering: Bitna is encouraged to continue speaking in spite of Salomé's intense pain.

Philosophical and Sociopolitical Readings of *Bitna, sous le ciel de Séoul*

(a) The critique of Western humanism

In the first instance, 'The Story of M. Cho and the Pigeons' emerges as a powerful critique of the nature/culture divide that persists in contemporary capitalist and globalised societies. To quote Bailly:

> L'invention, sans fin contestée par la vie même, d'une frontière absolue entre les animaux et les hommes, est directement responsable de ce qu'il y a de plus pénible dans l'humanisme, dès lors peu distinct d'un narcissisme d'espèce que l'on voit rebondir au long des âges.[51]

49 Ancius Boethius, *The Consolation of Philosophy* (Penguin Classics, 1999). It was written in 525.

50 Michel de Montaigne, *Essais: 1589–1592*. See *The Complete Essays of Montaigne*, <https://www.penguin.co.uk>.

51 Bailly, *Le parti pris des animaux*, p. 9.

This critique is displayed in the attitudes of the city dwellers towards M.Cho and the pigeons. We learn that, as the caretaker of a twenty-storey block of flats in Séoul, M. Cho has been granted permission to take his pigeons onto the roof for fresh air but only on two conditions: 'Mais vous veillez bien à ce que vos oiseaux ne fassent pas de saletés, et vous ne pouvez pas les mettre dans l'ascenseur' (p. 23).[52] If M. Cho fails to conform, the manager will receive an official warning from the bank that owns the block of flats (p. 26). Le Clézio is condemning here the totalitarian power of the bank, of commercial interests as well as the central importance accorded material possessions and physical appearance in consumer societies. Indeed, we learn that M. Cho 'qui a passé l'âge de la retraite' needs to climb a twenty-storey building five times a day in order to provide fresh air for the pigeons (p. 26). This would suggest not only a lack of empathy but also extreme cruelty. As Lévi-Strauss has written, the same barrier that separates humans from animals separates humans from other humans.[53] There is clearly no question of sharing the same space, of 'un monde commun' in the words of Latour, let alone any issue of human–animal equality.[54]

This spatial exclusion has its origin in philosophical dualism, in the separation of thought from the senses and of language from the living, an approach that underpins the writings of the linguist Ferdinand de Saussure, whose theories are still prevalent today in Western and Westernised societies.

Indeed, the very opening pages of 'The Story of M. Cho and the Pigeons' address the question of animal intelligence. The fact that the birds in their excitement to fly forget the bars of their cages is interpreted by M. Cho's neighbours as evidence of their stupidity: 'Les pigeons, ce sont les animaux les plus bêtes de la nature' and 'Avez-vous vu la taille de leur cerveau?' (p. 22). As research in ethology

52 The same city/country opposition is introduced in the first chapter where Bitna describes her own life. Originating from the country and then moving to Séoul, she meets the same hostility from her aunt in the city.

53 Lévi-Strauss, *Structural Anthropology 2*, p. 41.

54 See Latour, *Politiques de la nature. Comment faire entrer les sciences en démocracie*, p. 41. See section titled 'Equality: Human and Non-human', in Chapter 1.

has shown, the intelligence of birds is not related to the size of their brains.[55]

Furthermore, on a later occasion, the neighbours query M. Cho's naming of the pigeons: 'Pourquoi vous donnez des noms à ces oiseaux? Est-ce que les pigeons savent leurs noms, ce ne sont pas des chiens, tout de même!' (p. 25). M. Cho retorts: 'Mais ils connaissent leurs noms, madame. Ils sont beaucoup plus intelligents que votre chien, si vous voulez mon opinion.' And in response to the question 'Qu'est-ce que vos pigeons ont de plus que mon chien?' M. Cho replies, 'Ils volent, madame.' In this resort to humour, Le Clézio is clearly criticising the preoccupation of scientists with grading the intelligence of animals, as indeed they do of humans, often with dire historical consequences.

As we have already seen, the inferior status accorded animals per se is challenged at all levels of the text. M. Cho does not seek to dominate his pigeons: he treats them as his equals, as his helpers. His behaviour suggests love and tenderness as well as fidelity, feelings that would be communicated to Salomé. Indeed, his approach would give poetic expression to Leopold's ethic of the Earth, to the principles of equality and solidarity amongst all living forms.[56]

(b) The critique of everyday life in the contemporary city[57]

Indeed, in contrast to earlier texts such as *L'inconnu sur la terre* (1978) and *Voyages de l'autre côté* (1975), Le Clézio's poetic writing in this book is anchored in 'le temps réel', that is, in everyday life in a contemporary

55 Philippe J. Dubois and Élise Rousseau, *Petite philosophie des oiseaux* (Paris: Éditions de La Martinière, 2018), p. 116.

56 See Leopold, *A Sand County Almanac*, Chapter 1, p. 11, Note 47.

57 Alain Suberchicot writes, 'Il y a donc chez Le Clézio une inscription de l'écologie dans le monde social et dans l'histoire, car il ne peut pas exister de thématique écologique dans le texte littéraire qui pourrait s'abstraire de ces questions, sauf à banaliser le propos.' See Alain Suberchicot: *Littérature et Environnement: Pour une Écocritique Comparée* (Paris: Honoré Champion, 2012), p. 90.

globalised city.[58] The stories of the birds are interwoven with those that explore the themes of human cruelty, egotism and indifference to suffering. Le Clézio states with a reference to Salomé: 'Le monde qui est cruel et égoïste, parce qu'il ne s'occupe pas d'elle' (p. 201). Indeed, the book contains stories that present a powerful critique of contemporary society, a portrayal of the sexual, commercial exploitation of women and of male predatory behaviour, a manifestation of Le Clézio's own acute social awareness.[59] For instance, Bitna describes how on numerous occasions, she is followed by a stalker in the street, who then hovers outside her window, prompting her finally to seek new accommodation (pp. 103–105). Le Clézio succeeds here in conveying the horror of this experience, one that is shared by many women in the city. However, instead of being deterred by these stories and in spite of her increasing pain, Salomé urges Bitna to continue: like Faust, she is on a journey of exploration, of lived experience:

> Pour elle, je le sais, les histoires ne sont pas seulement des histoires, ce sont aussi des sensations qui l'effleurent, qui brûlent sa peau, des coups d'aiguille dans ses jointures, des vagues lancinantes derrière ses yeux. Elle les demande, et elle en a mal, elle les craint. (p. 156)

In another very moving story, Bitna relates the death of a female singer, Nabi, who, after a string of commercial successes, is replaced by a younger and cheaper version of herself, the trigger for her suicide (pp. 169–171). This would again suggest the cruelty inherent within the capitalist system, where money/profit is the ultimate value.[60]

58 In an interview with Lu Zwang, Le Clézio expresses his admiration for those philosophers 'qui ont essayé d'exprimer la philosophie à travers la vie quotidienne, les problèmes politiques'. He quotes Sartre, Camus and Badiou as examples. See *Les Cahiers J.-M. G. Le Clézio*, no 8, p. 99.

59 Women have a strong presence in almost all of Le Clézio's novels and short stories.

60 Seoul is one of the worst examples of neoliberal capitalism and social inequality. Referring to South Korea, Le Clézio states: 'Ce n'est pas un pays démocratique, c'est un pays ouvert à l'économie de marché, où règne le libéralisme mais pas la liberté. Une société archaïque, violente, étouffante, très raciste.' See L'OBS/N 2797-14/06/2018.

Furthermore, it is at this point in the text that Salomé informs Bitna of the suicide of her own parents. It is as if storytelling, identification with others, can have a curative effect, promoting a greater understanding of the suffering of others and of one's own. Salomé has come to a broader philosophical understanding, an acceptance of the everyday, of concrete lived reality, calling to mind the writings of Montaigne.[61]

At the same time, Bitna too feels that her storytelling has enabled her to arrive at a better understanding of life. Like Salomé, she has been transformed: indeed, she has become the messenger of a higher order. Just before the final story, that of the dying bird,[62] she states:

> Il me semble que tout a été ordonnancé pour accomplir cette histoire, que j'ai été en quelque sorte la messagère d'un ordre supérieur, céleste, et qu'après cela je ne pourrai plus jamais être la même personne. (p. 199)

Indeed, this intriguing sentence may suggest Le Clézio's own experience as a writer.

(c) *The critique of war*

'The Story of M. Cho and the Pigeons', arguably the richest in the book, is set against the historical background of the war between North and South Korea (1950–1953).[63] However, in spite of the war being technically over, there are still skirmishes at the border, and 'Tout cela pouvait recommencer n'importe quand' (p. 118). As we learn in the opening pages of the story, M. Cho, who comes from a village on the border with North Korea, is raising his pigeons so that they may one day bring

61 Michel de Montaigne (1533–1592), *Essais*, 1589–1592.

62 This is the story of a bird named O'Jay who, after falling from a tree, is nursed by a young girl, Naomi. When Naomi realises that the bird has only a few days to live, she comforts her by playing her recorded bird sounds. It is the most beautiful story in the book. As Bitna informs us, it is also a true story. Le Clézio as a child looked after a young jerboa. See *Les Cahiers J.-M.G. Le Clézio* no 14/15/ 2021–2022, p. 183.

63 There were approximately three million war fatalities and a larger proportional civilian death toll than in the Second World War. See Korean War, *Wikipedia*.

messages to the members of his family who have been separated by war (pp. 23–24). Having initially encouraged them to fly, he then organises 'un entraînement spécial pour ses oiseaux' to familiarise them with the sounds of shots (p. 118). They will also need to learn how to fly over the forest and the high mountains, 'des obstacles infranchissables à quiconque ne savait pas bien voler' (p. 95). As we already saw, the final flights of the birds mirror very closely those of the birds over the city; they are the expression par excellence of primordial energy and indeed of beauty (p. 123). Their successful arrival at the family farm would suggest the importance of repetition, the very act of poetic expression becoming a tool for combating violence and destruction.

M. Cho has indeed fulfilled his dreams. However, Bitna, the storyteller, is less optimistic about her ending:

> Mais ici, mais pour nous autres qui vivons ailleurs, rien n'est vraiment achevé. Le bonheur n'existe pas. Juste quelques rêves, quelques paroles. Juste le vent de la mer qui bouscule les plumes des oiseaux pendant qu'ils traversent l'estuaire.
>
> Et la réalité assassine.[64]

(d) Interculturality

At the same time, however, the story can be viewed as the expression of the utopian thought of Le Clézio. Indeed, the condemnation of war and the quest for a better world underpins all his writing.[65] As we have seen, far from being disillusioned by contemporary events, M. Cho strives on a personal level to fulfil his dream of bringing together members of his family separated by a political border. In the chapter titled 'Fin de l'Histoire de M. Cho pour Salomé, Fin Août 2016', we learn of the successful arrival of the pigeons with their messages at M. Cho's birthplace in North Korea, at 'l'autre coté de l'estuaire de la rivière', and that 'dans leur

64 p. 127. Recent years have seen a rise in tensions between North and South Korea. See *Le Monde,* 17 juin 2020.

65 Le Clézio has stated in an interview that his very early experience of war was the trigger for all his writing. See L'OBS/No. 2890–26/03/2020.

gorge commence le roucoulement de la paix' (pp. 124–125). They have reached the end of their journey: 'Ils marchent sur la terre, au milieu des habitants de la grande ferme collective des poiriers. C'est la fin de leur voyage' (p. 126).

Le Clézio could be deemed here to be promoting intercultural relations[66] between two countries or societies with very different political systems: communism and capitalism. Significantly, earlier in 2012, Le Clézio and his Mauritian colleague Issa Asgarally created La Fondation pour L'Interculturel et La Paix. As Le Clézio explains: 'L'interculturalité consiste en des cultures distinctes qui cherchent le dialogue l'une avec l'autre. L'une ne cherche pas à dominer l'autre.'[67] This statement may also call to mind Édouard Glissant's concept of Relation to be explored in the following chapter.[68]

66 See Thierry Léger: 'L'Asie et l'interculturel dans l'œuvre de Le Clézio', *Les Cahiers J.-M.G. Le Clézio* no 13/2020, pp. 119–132.

67 See Martha van der Drift, 'Entretien avec J.M.G. Le Clézio', *Les Cahiers J.-M. G. Le Clézio* no 5/2012, p. 162. See also Le Clézio's interview (France Inter) in 2017 where he states that the desire of South Korea for reunion with the North met with strong opposition from the United States. He also expresses his aversion for the very concept of a frontier.

68 See J. M. G. Le Clézio, 'Je me souviens d'Édouard Glissant', *La Nouvelle Revue Francaise*, No. 648 (2021).

CHAPTER 4

The Story of Dodo

This chapter is devoted to Le Clézio's book *Alma*, published in 2017 and set in contemporary Mauritius and France.[1] There are two intersecting voices, that of the Frenchman Jérémie Felsen, who visits the island, and that of the Creole Dodo, the primary object of this study.[2] Dodo's journey will take him from Mauritius to Paris and finally to Nice.[3]

My first section explores the Creole ecological vision as it emerges in the text and includes a focus on the intercultural dimension in a creative rendering of Glissant's concept of Relation. This is followed by a socio-political reading of this vision as it comes into conflict with ecological destruction and with contemporary forms of racism, social exclusion and violence both in Mauritius and in Europe. *The Story of Dodo* – largely a monologue – is further enriched when placed within the context of Jérémie Felsen's account, that is, against the background of the contemporary neo-colonial, consumer and racist society of Mauritius with its roots in slavery.

Significantly, both the names Dodo and dodo appear in the book.

As Jérémie Felsen informs us, the name dodo refers to the large birds endemic to the island of Mauritius that could walk but not fly and that were slaughtered during the colonial era.[4]

Indeed, his initial reason for visiting the island was the discovery of the gullet of a dodo in his father's collection. The Creole of the intersecting

1 Alma is the name of a family property in Mauritius originally owned by Le Clézio's ancestors. It is named after the wife of its first occupant.

2 The two are related. Dodo is also the name of the last descendant of the Felsens.

3 *The Story of Dodo* resonates with many of the author's earlier texts: Le Clézio has created a vast musical score, a web of sounds or notes.

4 Jérémie Felsen describes the capture and death of the last dodo on a ship bound for England. Its remains were sent to the Ashmoleum Museum in Oxford but were eventually destroyed (pp. 280–292).

narrative – Dodo – can be viewed both as a reincarnation of this original bird and as a contemporary rendering of the maroon, the escaped slave: indeed, we learn that it was at Port Louis, the historical centre of the slave trade, that Dodo caught an 'illness' – leprosy – destroying his face and making him a social outcast, the target of racism and of violence.[5] Significantly, in the very choice of the name Dodo, Le Clézio is also calling into question the nature/culture, human–animal divide at the core of Western philosophy and political power, the text resonating at this juncture with Le Clézio's earlier work *Haï*. Here Dodo's narrative trajectory can be viewed as a series of becomings, of becoming-bird: he embodies the freedom of movement and relationship to the Outside and natural world of the original birds that could walk as well as the dangers that confronted them. As we shall see, it is an assertion of freedom against all odds, an expression of the Creole vision and of the liberating power of the imagination. At the same time, *Alma* emerges as the most overtly political book that Le Clézio has written since his novel *Révolutions* (2008), addressing a number of contemporary issues with a degree of urgency.

Unusually, *The Story of Dodo*[6] is composed entirely in the present tense, collapsing the oppositions between the real and the imaginary as well as those between memory and forgetting: nothing disappears. As Le Clézio declared earlier: 'Pourquoi ne pas faire de l'ici, du présent, du déployé, notre vraie demeure'?[7] The register of language remains throughout oral: it is the intensity of sound that dictates meaning. Indeed, Le Clézio's writing is heavily influenced by the techniques and structures of the Creole storyteller and is almost totally devoid of abstraction or generalisation. Language

5 The portrait made of him at Port Louis was 'plutôt en noir' (p. 164). The word 'dodo' can refer not only to a bird but also to a black person, a tramp and a social outcast. There is no fixed identity: the person who gave him the 'illness' comments: 'Toi, tu es un drôle d'oiseau' (p. 29). Left without a nose or eyebrows, he looks like a bird. Indeed, on several occasions, he is addressed as such. The original dodo was descended from pigeons.

6 The whole story is produced in italics, giving the impression of a mosaic.

7 Le Clézio, *L'Extase matérielle*, pp. 186–187. Glissant writes that 'la poétique créole renforce en volume ou en mystère la présence de l'ici'. See Édouard Glissant, *La Cohée du Lamentin* (Paris: Éditions Gallimard, 2005) p. 32.

becomes alive, a living being: 'La langue que je parle, que j'écris, je la sens plutôt comme un être vivant qui bouge, qui change, qui s'enfuit. Un flux et un reflux de paroles.'[8] It is syntax itself that liberates the expressive poetic power of langage. As Deleuze asserts: 'La syntaxe est l'ensemble des détours nécessaires chaque fois créés pour révéler la vie dans les choses.'[9] In a sustained focus on orality, on sound and repetition, the text differs in style from any previous work of Le Clézio.

After an introduction to the Creole ecological vision, this chapter will present Dodo's story as a journey through space and time in which he will make a number of significant encounters.

The Creole Ecological Vision

In the opening pages of Dodo's account headed 'Mon Nom est Dodo', the reader is introduced to the Creole vision of the world originally associated with the escaped slaves of Mauritius. We learn that his Creole nurse, Artémisia, 'connait toutes les devinettes' (p. 18), an intertextual reference to the earlier mentioned book *Sirandanes*. Two key passages in the preface give expression to a poetics of the Caribbean landscape and are invaluable for an understanding of the narrator's project and of the workings of Relation. The first posits the interconnectivity of all forms of life:

> La vie: un regard neuf sur le monde, sur les êtres et les choses. L'univers des sirandanes est un lieu sans frontière, où nul n'est séparé. Les végétaux, les animaux, les hommes et les éléments sont encore très proches les uns des autres, comme au premier moment de la création. (p. 14)

A reference a few lines later to 'la grand-maman araignée' anticipates the human–animal, self–world relationship embodied by Dodo, a contemporary rendering of the bird who can walk. It would directly address the Deleuzian theme of becoming.

8 *Sirandanes*, p. 63.

9 Deleuze, *Critique et clinique*, p. 12.

At the same time, the opening pages of *Sirandanes* can be viewed as an introduction to Le Clézio's concept of the sacred. Here he develops the themes of a universal language and of a primordial energy introduced in *Haï*:

> On voudrait parler d'animisme. C'est la leçon philosophique des sirandanes, cet art de la parole si léger et si grave. Il y a, je crois, un message étrange qui est venu de la 'grand-terre', ou peut-être même du cœur de l'Afrique, et qui a gardé la vérité profonde des religions et des mythologies premières, qui a conservé cette connivence entre les hommes et leur monde, ce lien qui unit les premiers chasseurs et les premiers collecteurs à la savane et à la forêt. On sent ici la force des éléments, le ciel, les orages, les vents, la puissance de la vie dans tous ses dessins, dans tous ses gestes [...]. (p. 15)[10]

This ecological vision finds creative expression in the narrative trajectory of Dodo: he is constantly on the move and exposed directly to the cosmic forces of rain, wind and sun. The very act of walking, of living Outside, would suggest the theme of errantry. As Dodo declares in the opening chapter, 'Tous les jours je marche' (p. 19) and later, 'Moi je sais marcher, c'est ce que je fais de mieux' (p. 248). Indeed, he is not constrained by any geographical borders: the text is scattered with numerous place names, rendering concrete the notion of a journey as well as blurring the lines between fiction and 'le réel'. For instance, his travels around Mauritius include Minissey (p. 20), Valeta (p. 20), Beau Basin (pp. 46, 49), Rose Hill (pp. 46, 49, 149), La Louise (pp. 49–50, 50–51, 53), Bonne Terre (p. 62), Crève Cœur (pp. 64–65, 167), Curepipe (p. 52), Caudan (p. 149), and Port Louis (pp. 28, 49, 164) as well as two cemeteries, 'le cimetière de l'Ouest' (pp. 127–129) and 'le cimetière Saint-Jean' (pp. 148, 167).The text is indeed striking for its geographical precision, drawing the reader's attention to the evocative power of naming.[11]

10 A parallel can be established with Glissant's concept of the Tout-Monde (Relation) and of the sacred: 'Le Sacré procède peut-être pour nous de cette Relation, non plus d'une Révélation ni d'une Loi.' See Édouard Glissant, *Traité Du Tout-Monde* (Paris: Éditions Gallimard, 1997), p. 113.

11 Several of these place names are also associated with slavery, e.g. Port Louis (p. 502), Minissy (pp. 449–450), and Crève-Cœur (p. 453). They would bring to mind the historical image of the bird that could walk.

As I shall show, Dodo's journey will be one of flight, establishing multiple relations between different spaces, cultures and forms of life, and adapting to each situation as it arises. It would give expression to Glissant's concept of Relation as movement, as a vitalist force and one of constant renewal.[12] Indeed, on several occasions, Dodo is likened to a bird, for Glissant the very image of creolisation.[13] His Creole friend, Ayeesha, addresses him at one point: 'Hé-hon! Dodo, Dodo-bird! Kot to été?' (p. 23). Later in Paris, the refugees sing his name: 'Dodo! Dodoo!' and again 'Dodo! Doo-o-oh!' (p. 256). As we learn from Jérémie Felsen's account, this was the cry of the original birds that could walk on the island.[14] In addition, on his journey from Mauritius to Nice, Dodo will never be without a travel bag imprinted with the name of a bird, 'kestrel'. Indeed, his trajectory is one of constant deterritorialisation, and flight itself would suggest the freedom of the imagination. Moreover, as stated earlier, the language of Dodo, in contrast to that of Jérémie Felsen, is characterised throughout by poetic sensorial expression, a language closer to that of birds.

I shall begin by focusing on the island of Mauritius and on Dodo's Relation to nature and to the elements. As Glissant writes: 'Qu'est-ce, un pays, sinon la nécessité enracinée de la relation au monde?'[15]

Mauritius: Relation and the Natural World

The opening chapter of *The Story of Dodo*, headed 'Mon Nom est Dodo', alerts the reader to the central role played by primordial forces, those of the sky and the earth.

12 Édouard Glissant, *Poétique de la Relation* (Paris: Éditions Gallimard, 1990). See 'La Relation est mouvement' (p. 186), 'La Relation relie (relaie), relate' (p. 187), and 'la relation, qui est nouveauté du monde' (p. 191).

13 Glissant, *La Cohée du Lamentin*, pp. 1–18, 33, 73, 74.

14 See the chapter titled 'La Mare aux Songes', p. 40.

15 Édouard Glissant, *L'Intention Poétique* (Paris: Éditions du Seuil, 1969), p. 72.

Sky

Here Dodo describes his journey towards the Saint-Jean cemetery in a seven-line sentence: each moment is, as it were, an awakening to the world, a series of beginnings, drawing immediate attention to the poetic, expressive function of the Creole language. The focus here is on the relationship between sky and earth:

> Je pars des champs de cannes, je marche au soleil par les petits chemins, la terre est rouge et craquelée, je sens la brûlure sur mon visage et sur mes mains, et quand je traverse les routes du côté d'Ébène, les nuages s'amoncellent au-dessus de la montagne, de grands nuages blanc et noir qui s'entrechoquent, je sens le vent froid de la pluie. (p. 22)

The rhythm of this sentence conveys both the movements of Dodo and the transformation of the landscape, from sun and heat to cold and rain. Sensorial intensity is rendered in the minute details of the description, in the accumulation of verbs and in a process of repetition, all distinctive features of the Creole language that will come to characterise the text as a whole.[16] Indeed, we may note the central role played by the very sound of words. The following semantic fields emerge and will dominate the landscape of Mauritius:

> Cosmic forces: le soleil, la terre, la pluie.
> Temperature: je sens la brûlure du soleil, je sens le vent froid
> Bodily parts: sur mon visage, sur mes mains
> Size: les petits chemins, de grands nuages
> Colour: La terre est rouge, les nuages blancs
> Movement: je pars, je marche, je traverse, les nuages s'amoncellent.
> Sound: la terre (...) est craquelée, les nuages (...) s'entrechoquent.

Here Dodo is clearly undergoing a fusional experience with the elemental forces and with the cosmic. As Frédéric Gros writes: 'Le marcheur s'accomplit dans l'abîme d'une fusion, le flâneur dans l'explosion d'une

16 Glissant speaks of 'le premier geste de l'art, l'effusion magnétique, la tentative de fusion avec les énergies du monde'. See Édouard Glissant, *Une Nouvelle Région du Monde* (Paris: Éditions Gallimard, 2006), p. 74.

dispersion indéfinie d'éclats.'[17] The text is highlighting the performative function of language, the simultaneous emergence of subject and world, or 'la co-naissance au monde et de soi-même', in the words of Paul Claudel.[18] At the same time, it is an experience of the spiritual within the natural world, of the invisible within the visible. Of particular significance here is the repeated image of the clouds, one that will later dominate the landscape of Mauritius and become a figure of Relation: it is their movement that will mirror Dodo's own journey back to the space of origins, that is, to Alma to 'un temps originel', in the words of Deleuze.[19] The text resonates here with the thought of Bachelard, for whom the very image of the clouds would suggest a relationship to the infinite, an expression of the material imagination.[20]

Importantly, the story goes on to contrast this experience, 'le vent froid de la nuit', with the reaction of the people around him: 'Les gens se hâtent, penchés sous leurs parapluies, les filles des collèges s'agrippent aux bus, leurs voix crient des 'ah et des oh' (p. 22). The image here of people rushing indoors would suggest a critique of Western dualism, of the oppositions nature/culture, sky/earth, infinite/finite underpinning Western thought. The sense of shelter becomes a withdrawal within oneself, within human constructs. For Dodo, the explorer, on the other hand, life itself becomes a journey, a process of constant transformation and renewal. Like a bird, he is always on the alert. As Le Clézio writes, 'Voyager, c'est avoir les yeux ouverts quand tout le monde dort' (p. 179).[21]

By focusing on the weather in the opening stages of the journey, Le Clézio is calling into question deep-seated prevailing value systems: the centrality of the human remains at the core of mainstream Western philosophy, contrasting sharply not only with the Creole vision but also with

17 Frédéric Gros, *Marcher, une philosophie* (Paris: Flammarion, 2011), pp. 242–243.

18 Glissant refers to this statement by Claudel in *L'Intention Poétique*, p. 144.

19 Gilles Deleuze, *Proust et les signes* (Paris: Presses Universitaires de France, 1964), p. 26.

20 Gaston Bachelard, *L'Air et les Songes. Essai sur l'imagination du mouvement*, 1943 (Paris: José Corti, 1994), p. 14.

21 We learn that Dodo cannot sleep; he lives in an eternal present: 'c'est que ma journée n'en finit pas, c'est une route sans fin' (p. 62).

Zen Buddhism, Taoism and shamanism, philosophies with which Le Clézio has earlier engaged.[22]

Sky and earth: The cemetery

In contrast to those sheltering under their umbrellas, Dodo, on his way to a cemetery, is described as walking towards the rain 'qui tombe et coule sur mes joues, qui mouille ma chemise et descend le long de mes jambes' (p. 23). The rain/water, an elemental force of pure energy, is presented here as a means of communication not only between human and non-human but also between the living and the dead, a beautiful rendering of the poetics of Relation: 'Moi j'aime la pluie au cimetière Saint-Jean. Papa, Maman, vous aimez bien la pluie aussi, quand on est mort on aime la pluie parce que ça ressemble aux larmes' (p. 23). Through the activity of the imagination, there is an expansion of consciousness, a Proustian association between sensation and memory, between the material world and immaterial time.[23] Speaking to the dead here is clearly an expression of generational memory. It would also point to Bergson's 'élan vital', the belief that all matter is alive, that all forms of life are conscious, are intelligent.

In a second visit to a cemetery, 'le cimetière de l'Ouest', Dodo, upon discovering the tomb of his ancestor – the first Felsen who came to Mauritius from France – lies down on the earth in a communication with the dead, the text highlighting the relational importance of touch. In an experience of metamorphosis, of 'becoming-bird', Dodo now feels himself flying in the direction from which his ancestor came: 'Je glisse avec les nuages, dans le ciel ils ne dorment jamais, eux non plus, ils avancent en plein ciel, et moi j'avance avec eux, je vais de l'autre côté de la mer (p. 128).[24] Indeed, he is now envisaging the end of his own journey, a return to the space of origins,

22 See, for example, *Haï* (1970), *Désert* (1980), and *La quarantaine* (Paris: Éditions Gallimard, 1995).

23 It is not rational time that restores the past. As in the case of Marcel Proust, sensation and memory are linked.

24 The text refers here to Axel Felsen, an ancestor of Jérémie Felsen and Dodo Fe'sen who lived in nineteenth-century colonial Mauritius.

to the 'other side': 'Un jour si je suis capable, je voyage là-bas, au pays des Fe'sen. C'est un pays dans les nuages, un peu comme ici, au cimetière de l'Ouest' (p. 129). As in the case of the first visit to a cemetery, Le Clézio is collapsing the division not only between life and death, but also between past and future, a further expression of Creole consciousness.

The tree

This relationship between the living and the dead is also conveyed in the image of the tree, regarded as sacred in Creole thought, as 'l'événement continuel, infini, impensable du vivant' in the words of Patrick Chamoiseau.[25] In the chapter titled 'Alma', Dodo describes how he goes to Crève Cœur with presents for the Creole Yaya, a former slave, who is buried 'entre les racines du mangier' (p. 65). Then much later, when contemplating leaving Mauritius, he returns and again leaves presents 'entre les racines du mangier' (p. 167).[26] The repetition like a mantra of the image of the roots reminds us that the earth is alive, that it is conscious: it represents the sacred or pure energy, the space of origins. Intertextual links can be discerned with the writings of Bachelard, for whom the roots of a tree likewise represent a synthesis between life and death.[27]

Importantly, these rituals of memory in *Alma* become a reassertion of life, an attempt to combat an historical forgetting embedded within Western thought, within its linear concept of time. Dodo is aware that, once he has left the island, no one will visit Yaya's grave. Furthermore, as the Creole Aditi informs us, most of the trees on the island have already

25 See Patrick Chamoiseau, *Les Neuf Consciences du Malfini* (Paris: Éditions Gallimard, 2009), p. 221.

26 When Dodo was young, Yaya built a house under the mango tree for her ancestors from Africa, for 'grand maman araignée' et 'grand-papa soursouris' (p. 63), a quotation from *Sirandanes*.

27 Gaston Bachelard, *La Terre et les Rêveries du repos* (Paris: Librairie José Corti, 1948). Referring to a text by Paul Valéry, he writes: 'Du côté des racines, on rêve bientôt la Terre entière comme si elle était un nœud de racines, comme si seules les racines pouvaient assurer la synthèse de la Terre' (p. 311).

been destroyed.[28] The concept of the sacred has become increasingly fragile in contemporary Western or globalised societies. Humans, through a lack of imagination, can indeed jeopardise life on earth, as Dodo will sense on his later arrival in Paris.

Relation and Interculturality: Religions

'Tous les jours, Je marche. Tout le jour', declares Dodo (p. 19). Walking around Mauritius involves a number of encounters with people or places representing different religious beliefs or cultures, giving further expression to the notion of Relation as conceived by Deleuze and Glissant, that is, as a movement in which 'on entre dans les zones de voisinage, plutôt qu'on n'acquiert des caractères formels'.[29] This movement, one of multiple connections, of interlinked relationships, is conveyed initially in the text through the very use of naming, an address to the reader's imagination and to the poetic, associative power of language.

The Swedenborg church

In the opening chapter of the book, we learn of Dodo's visit to the Swedenborg church, motivated by the need to acquire another pair of shoes. Significantly, Le Clézio draws the reader's immediate attention to the geographical location of the building, the rhythm of the sentence conveying the movements and direction of the walker: 'Je vais vers les hauts, du côté du Trou aux Cerfs, du côté du Jardin botanique, de l'église

28 As we learn from the interwoven account of Jérémie Felsen, Aditi runs a wild life foundation. See p. 81.

29 This is a quotation from Deleuze appearing in Glissant's *La Cohée du Lamartin*, p. 136. In his concept of Relation, Glissant draws upon the Deleuzian notion of the rhizome, that of multiple connections, of interlinked relationships. See Deleuze and Guattari, *Mille Plateaux*, pp. 9–37.

swedenborgienne' (p. 19).[30] The place names here together with a mention of 'les hauts' – the forest and mountains – would suggest a relationship with the non-human world, calling to mind passages from the text *Sirandanes*, the embodiment, as we saw, of the Creole ecological vision. It is also the landscape of the dodos, of the original birds who could walk.

In addition, Dodo's reception by the staff is one of kindness and hospitality, an unconditional acceptance of cultural difference: 'Les nénénes me donnent les souliers avec un mot gentil' (p. 20). Moreover, they address him with the neutral term Dodo, and not by his French name, Fe'sen Coup de ros, eschewing any single cultural or family affiliation, any fixed identity.

The intercultural is further conveyed in the list of names in the newspaper in which the shoes were wrapped: mainly Chinese or Japanese with the occasional mention of a French or English first name. This image would acquire a wider political import against the background of the wealthy class-driven, and racist society of Mauritius described in the interwoven account of Jérémie Felsen and whose negative impact will later be experienced by Dodo himself.[31]

Importantly, the very choice of the name, Swedenborg, would illustrate the poetic power attributed to naming in the book, a direct address to the reader's imagination.[32] As in the case of place names, we are sent on a quest: Emanuel Swedenborg was a Swedish philosopher, Christian mystic and scientist inspired by Plotinus[33] who, like the Creoles, attached key importance to the relationship of the human to the non-human world, and indeed wrote extensively on Cosmic Consciousness, on Light and on

30 Le Clézio is awakening the reader's curiosity, blurring the lines between fiction and geographical reality, addressing the mapping skills of the reader. For instance, the Swedenborg Church ('l'église swedenborgienne') is at Curepipe, and 'Trou aux Cerfs' is just north of the Botanical Gardens (p. 19).

31 See, for example, the chapter titled 'Pomponette' (pp. 143–145).

32 In his Nobel Prize speech, Le Clézio expresses his admiration for 'l'architecture visionnaire d'Emanuel Swedenborg'. See 'Conférence Nobel, le 7 décembre 2008'.

33 Plotinus saw music, metaphysics and love as manifesting the truth of absolute, infinite beauty. He also viewed atoms as knots of energy, anticipating modern science and the theories of matter of Bergson. His work on light had a profound influence on Albert Camus.

Beauty.[34] As we have already seen, the opening stages of Dodo's journey focus on a fusional relationship with the sky and earth. Beauty itself will emerge as a central ingredient of Relation in the narrator's journey around Mauritius.

At the same time, a common ground is conveyed on the political level: throughout his life, Swedenborg championed the cause of love and world peace, adopting a strong ethical and political position, one that will later mirrored in the text as Dodo's travels take him to France.

Père Chausson

The concept of interculturality, or Relation is further explored in the chapter headed 'Un lavement de pieds' (pp. 149–156) in a poetic rendering of the image of water associated by Christian and Creole alike with the renewal of life and with the sacred. Here Le Clézio describes Dodo's visit to the Roman Catholic cathedral at Caudan in Mauritius and his participation in a ceremony organised by the priest, Père Chausson, where the tramps are having their feet washed. Significantly, we learn from the outset that Père Chausson also works in the north of Mauritius, in the church of Cap Malheureux, where he conducts marriage ceremonies between Creole women and Muslims. As Dodo comments, 'c'est pour ça qu'il a cette chasuble d'un côté blanche avec la croix de Jésus, de l'autre verte avec la lune de Mahomet' (p. 153). Indeed, read in the light of Jérémie Felsen's account, this episode presents a stark contrast to the religious intolerance of the white middle class on the island (pp. 143–144). It could also be viewed within the wider context of Le Clézio's critique of the language of abstraction and of generalisation associated traditionally with the Christian church. As we shall see, the episode is distinguished by

34 Emanuel Swedenborg (1688–1772) was interested in the relationship between the natural world and the spiritual, particularly in the movement of transcendence. See also Gaston Bachelard, *Le Droit de rêver* (Paris: Presses Universitaires de France, 1973), pp. 128–132. Here Bachelard describes Swedenborg's images as 'la sollicitation d'un mouvement ascensionnel, comme une invitation constante à un avenir aérien' (p. 129).

its highly poetic register displayed in the choice of Biblical passages as well as in the use of associative thinking, a direct address to the imagination.

Père Chausson opens the ceremony of the washing of feet by establishing a parallel between the tramps of Mauritius and Jesus's disciples, by creating as it were a new imaginary or a series of becomings: 'il dit que ce jour est important parce que Jésus-Christ est avec nous, et que lui aussi lave les pieds le jour du jeudi saint, avant d'être crucifié' (p. 154). The identification here with the disciples would enhance the tramps' image of themselves and convey a sense of personal value in a context of social alienation and extreme hostility.

This introduction is followed by a girl giving a reading from St John 13:1. The passage is an expression of love, again an experience not easily encountered in a society hostile towards the homeless:

> Avant la fête de la Pâque, Jésus, sachant que son heure était venue de passer de ce monde à celui de son père, après avoir aimé les siens dans ce monde, les aima jusqu'à la fin. (p. 154)

The use of repetition here 'aimé, aima', conveys the intensity and duration of Jesus's love.

Indeed, the girl's voice is described as 'claire' and as 'une voix comme l'eau qui coule de l'arrosoir', the very sound of the word 'coule' suggesting the act of flowing (p. 154). Le Clézio introduces at this point the theme of beauty associated earlier with Swedenborg: 'elle a une voix aigue qui tremble un peu [...] et elle trouve ce passage très beau'. Indeed, it is the very beauty of sound that has a transformative impact on the tramps: 'Les clochards ont arrêté de ricaner, il y en a même un qui se met à pleurer' (p. 154).

The girl continues reading from the Bible, focusing now on the act of washing, the image of water evoking the renewal of life in both Creole and Christian thought: 'Puis il versa de l'eau dans le bassin et se met à laver les pieds des disciples, et il les essuya avec le linge dont il était ceint' (p. 155).[35] The effect is magical: the tramps, hypnotised, now look down at their own feet.

35 John 13:5. All the quotations are from the Revised Standard Version of the Bible.

The parallel with the biblical is sustained. A further quotation presents the refusal of the disciple Simon Peter to have his feet washed, and the response of Jesus: 'Ce que je fais, tu ne le comprends pas maintenant, mais tu le comprendras bientôt' (ibid.).[36] When some of the tramps start to imitate Simon Peter, Père Chausson steps in to save the situation: by opening his green and white chasuble 'comme des ailes d'un oiseau', he indicates that he is welcoming them all, caring for them all, Creole and Christian alike (p. 156). The spreading of the chasuble now produces the intended transformation:

> 'Non', dit Pierre. 'Jamais vous ne me laverez les pieds.'[37] Et Jésus lui répondit: 'Si je ne te lave pas, tu ne pourras pas vivre avec moi.' Alors Simon-Pierre dit: 'Bien, Seigneur, lavez-moi les pieds, et aussi les mains et les cheveux!'[38]

The text plays here on the double meaning, spiritual and everyday, of 'vivre avec moi'. It would suggest the healing, transfiguring power of the imagination, of the poetic. At the end of the ceremony, Dodo too feels cleansed, transformed, 'et j'ai les pieds très propres, moi aussi'. Le Clézio is drawing attention to the philosophical and poetic strength of water as an agent of renewal of life and of metamorphosis. The text may call to mind the image of the rain explored in the initial stage of the narrator's journey around Mauritius.

Interestingly, although the dominant image has clearly been that of water, there are also references to the sun, reinforcing the Creole component. Père Chausson is described as giving his opening speech with 'le dos au soleil' (p. 154), and the girl, reading from the Bible, 'se place face à nous avec le soleil dans le dos' (ibid.). All in all, the chapter can be viewed as a powerful illustration of the poetic function of language, of the recourse to the senses and the imagination as a vehicle for transcending cultural and political barriers. Indeed, the frequent references in the text to beauty and the very choice of the biblical passages with their focus on the power of sound would call to mind the text *L'Intraitable beauté du monde* by Glissant

36 St John Gospel, verse 7.

37 Ibid., verse 8.

38 Ibid., verse 9.

and Chamoiseau where beauty is presented as an essential ingredient of Relation.[39] At the same time, the passage highlights Le Clézio's concept of the sacred: it is not a revelation as in traditional Christianity but the very act of creation of life.

Relation and Interculturality: People

Vicky

A central role in intercultural relations is also played by the English nurse Vicky, whom the narrator met in hospital after being seriously injured by the students in a graveyard (p. 130).[40] It is this friendship that will facilitate Dodo's journey of discovery, his eventual departure for Europe. Indeed, Vicky shares many of his bird-like qualities. The scene opens by foregrounding the value of beauty associated, as we have seen, with the ceremony conducted by Père Chausson. Here it is expressed initially in terms of vision and of colour: 'L'infirmière est très belle, elle est grande avec des cheveux blonds et des yeux bleus' (p. 132). She is also 'la plus belle fille de l'hôpital' (p. 133).

This concept of Relation is further explored in a second meeting, this time outside the church, Marie Reine de la Paix, just prior to Dodo's departure for Paris.[41] Here Dodo voices a strong desire for physical contact expressed in images of the skin and in notations of touch and smell: 'sa peau est couverte de petits grains de beauté, j'ai envie de toucher sa peau pour sentir sur ses joues le duvet de fruit, j'ai envie de l'embrasser pour sentir son

39 Édouard Glissant and Patrick Chamoiseau, *L'Intraitable beauté du monde* (Paris: Galaade, 2009). See also www.Galaade.com.

40 For Creole populations, the graveyard is a sacred space. The behaviour of the students would suggest Western attitudes to the dead, attitudes that are grounded in the life/death duality. The students are privileged Carnegie students from a variety of countries.

41 The reference here to a Catholic church also strengthens the theme of interculturality.

odeur de fruit' (p. 164). Intensification of desire is communicated here in the repetition of 'j'ai envie' and in the accumulation of physical details – a fusion of the poetic and the realistic. Indeed, the sentence conveys in all its immediacy the tragedy of Dodo's situation, his position as a social leper unable to enjoy 'normal' human relations, his longing to give full expression to his desires.

In a second reference to the same scene at the church, Le Clézio focuses on the relationship of the two to the cosmic forces: 'Alors nous restons là assis sur un banc devant Marie Reine de la Paix, avec le soleil qui va boire l'eau de la mer et le vent chaud qui souffle, ça ne peut pas finir' (p. 165). The rhythm of the sentence, the use of the conjunctions 'avec' and 'et' suggests harmony, a sense of pure presence and of timelessness. The passage concludes in a further awakening of Dodo's senses: he can feel 'l'odeur de sa peau et de ses cheveux blonds et regarder ses gros yeux bleus' (p. 166).

Later, the text returns to this scene at the church in a process of sensorial intensification: this time the sense of touch is brought into play: 'Elle m'embrasse pour la première fois, je sens l'odeur de violette dans son cou, dans ces cheveux' (p. 178). In the very image of the kiss, the figure of the nurse acquires enriched meaning: she is indeed communicating 'the breath of life'. Dodo has established a human relationship that is clearly much closer and more intense than in previous encounters.

As will also become apparent, both Dodo and Vicky embody the theme of flight, of 'becoming-bird'. Like birds, they enjoy freedom of movement around the island and are associated with the outside: Vicky's home is never mentioned, and the absence of a surname may point to a high degree of family and cultural independence, a lack of any identifiable roots. Furthermore, when Dodo is eventually disillusioned with life in Mauritius, it is Vicky who will provide him with a passport and the necessary contacts in Paris. She will also give him a blue travel bag with the name 'Kestrel'[42] written in white letters and 'avec le dessin d'un oiseau blanc' (p. 165). We also learn at this juncture that she has a baby, pointing to the act of renewal of life, of continuous creation.

42 'Kestrel' is the name of the company for which Vicky works.

At the same time, however, the choice of the place name where the two meet would suggest a broader political reading of the concept of Relation. There are three mentions of the church, Marie Reine de Paix, the name signalling peace in marked contrast to the social and racial violence that the narrator has encountered on the island (p. 132) and that he will later witness in Paris. Furthermore, in her function as nurse and healer, Vicky embodies the values of selfless love and care, values that transcend all national and political boundaries.

La fille aux cheveux bleus

In the chapter titled 'Lézard' describing Dodo's arrival in Paris, we learn of a meeting at a fairground with a homeless girl who will eventually follow Dodo to Nice. In stark contrast to the people he has already encountered, she is deaf and dumb and 'elle ne peut pas parler sauf avec ses doigts' (p. 246). This absence of speech, however, rather than being a deficiency or handicap, is presented as a source of strength: in a return to this episode in the chapter called 'Le prophète', Le Clézio writes, 'je l'aime bien parce qu'elle ne parle pas, seulement avec ses mains et avec ses yeux, et je suis content parce qu'il y a toujours trop de mots dans le monde' (pp. 258–259). Words alone, that is, representational language, can become a tool of oppression, of confinement within social and political expectations, the recipe for conflict as will be evidenced in Dodo's later experiences in Paris. Instead, the girl speaks in a body language, a richer, more expressive form of communication. Indeed, as Le Clezio wrote earlier: 'Il y a tant de choses qui passent par le silence' (p. 246). In the choice of a character conventionally termed 'disabled', the episode can be considered a striking example of Relation: as we shall see, it emerges as a poetic exploration of the power of beauty associated in the text with the figure of the artist. The episode contrasts with Dodo's earlier encounter with Père Chausson where beauty was associated predominantly with the sounds of language.

Indeed, what initially attracts Dodo's attention at the fairground and will become the primary catalyst for Relation is colour itself, the colour blue of the girl's hair. Their meeting opens with the sentence 'C'est ici que je rencontre la première fois la jeune fille aux cheveux bleus' (p. 246). A few lines later, sensorial intensity is conveyed in the accumulation of detail: like the Indigenous peoples in *Haï* who paint their body, the girl has become an artist:[43]

> J'aime bien ses yeux bleus, et aussi la couleur de ses cheveux, les petits cheveux à l'arrière sont noirs et devant elle a de grandes mèches peintes en couleur bleu, elle les attache quelquefois avec un chouchou. [44]

Indeed, the use of contrast here, of notations of light and dark colours together with the choice of the term 'peintes' would anticipate the painter whom Dodo will meet in Nice towards the end of his journey (p. 309).

Later in the text, in the chapter called 'Le prophète', Dodo returns to an evocation of the meeting at the fair, but this time the portrait is enhanced. The focus remains on colour, but there are variations in the detail suggesting a transformation. This is conveyed in the very rhythm of the sentence: the portrait is alive, animated:

> Maintenant elle est en robe longue et elle a la peau brune et des yeux clairs, elle a les cheveux teints en bleu mais la peinture s'en va et en dessous ses cheveux sont noirs. (p. 259)

Painting the body here is an expression of consciousness, of thought. As Le Clézio writes in *Haï,* 'Peinture vivante, peinture de la pensée sur la peau' (p. 134) and 'Chaque homme, chaque femme est un tableau. La peau est vivante' (p. 121). Le Clézio then goes on to enhance the portrait by introducing the act of touch:

> La jeune fille aux cheveux bleus pose sa tête sur mon épaule et j'aime sentir son poids, j'ai les yeux ouverts toute la nuit et j'écoute sa respiration, et quand c'est le matin, elle glisse sur le sol et elle dort pliée, la tête contre ma hanche. (p. 259)

43 As Le Clézio writes in *Haï*: 'Aussi, l'art indien n'est pas le secret de quelques artistes. Il est un langage que chacun possède et que chacun exprime' (p. 117).

44 The references to hair dye would break with more mythical presentations of the nomadic in Le Clézio's earlier texts.

The very reference to the act of breathing here may suggest a link between touch and the 'breath of life', between touch and *le vivant*.[45]

Again, as in a musical score, the chapter 'Vers le sud' returns to the same episode but with variations on the theme: there are noticeably more frequent references to touch conveying sensorial intensification:

> La jeune fille aux cheveux bleus [...], elle se couche près de moi et elle met sa tête sur mes genoux, je sens qu'elle dort et je caresse doucement ses cheveux bleus, ils sont raides, mais j'aime bien les toucher. Si j'ai mal au dos, je m'allonge par terre et la fille se colle contre ma poitrine [...]. Je sens la chaleur de son corps. (p. 296).

Colour – a quality that Dodo lacks – is clearly a communication of energy, of bodily heat.[46] At the same time, touch can be interpreted as the expression of love, heightened in the use of repetition: 'je caresse doucement ses cheveux bleus' and 'j'aime bien les toucher'.[47]

This episode could be said to carry a rich semantic investment. The colour blue – the girl's very name – would also suggest an identification with the cosmic, in other words, a mystical experience.[48] The two are both associated with the Outside, with the Open[49] and are undertaking a journey of discovery (p. 247). It is clearly movement itself that is valorised, not the goal: like Dodo and Vicky earlier, the girl is undergoing a metamorphosis, a 'becoming-bird':

45 I note in Chapter 2 the crucial role played by touch in the creation of meaning: like the spiders who have poor hearing, the girl is speaking with her whole body.

46 Intertextual links may be detected, as demonstrated by Michel Collot, who views painting as the expression of 'les énergies de l'univers'. See Michel Collot, *La Pensée-paysage* (Actes Sud/ ENSP, 2011), p. 241.

47 For Sigmund Freud, love is the expression of energy: 'Nous appelons ainsi l'énergie [...] ces pulsions qui ont à faire avec tout ce que l'on peut comprendre sous le nom d'amour'. See Jean-Yves Tadié, *Le Lac inconnu: Entre Proust et Freud* (Paris: Gallimard, 2012), p. 121.

48 The colour blue as well as the phrase 'le bleu du ciel' is associated with a mystical experience in several of Le Clézio's texts. See, for example, the earlier mentioned Bedouin nomads, 'les hommes bleus', in his novel *Désert*.

49 For Bailly, 'l'ouvert n'est que l'éternelle présentation au présent et il est, comme tel, sans passé et sans avenir'. See *Le Versant animal*, p. 40.

> Elle laisse sa famille et les Gitans de la porte de l'Est, elle marche avec nous, elle est pareille aux oiseaux qui suivent un bateau, c'est juste parce que nous allons quelque part, elle n'a pas besoin de savoir où elle va. (p. 294)

The power of this story – and its distinguishing factor – lies in the very close attention afforded bodily detail. Le Clézio is able to combine realism such as the reference to hair dye, with poetic expression, thereby collapsing the conventional rhetorical divisions that still persist in much of Western literature. Its impact on the reader lies in its very simplicity, in the expressive value of silence, of physical appearance and of gesture. Beauty can be found everywhere, as Glissant eloquently claims.[50] Indeed, Le Clézio is presenting the artist not in the context of an institution or indeed of a permanent residence, but as a 'voyageur' unencumbered by material possessions: 'Elle ne porte pas de bagages, elle n'a que ses habits, son jeans délavé et troué, son blouson en nylon et son foulard autour du cou' (p. 294).

Furthermore, in this image of the artist, Le Clézio is also dissolving temporal borders between past and present, calling into question the linear unfolding of time and space and highlighting the phenomenological role of the imagination, a theme that will be later developed in Dodo's encounter with an artist in Nice. As stated earlier, nothing disappears, nothing is lost. As we have seen, the text returns on two occasions to the same meeting at the fair. Each time, however, we are provided with more details, the portrait is refined. This process of repetition and accumulation is also one of sensorial intensification, a distinctive feature of the Creole language. In other words, like the repeated gesture of placing presents under the tree, art itself is an expression of the creation or resurrection of life.[51]

50 See Glissant, *Une Nouvelle Région du monde*, pp. 14–15.

51 See *The Story of Dodo*, pp. 166–167.

Relation and Music

It is music, a central feature of Creole culture,[52] that plays a key role throughout the novel in bringing together different spaces, cultures and times, giving further expression to the theme of flight, to Dodo's bird identity. Indeed, our immediate attention is drawn to the ecological function of sound and rhythm as attributes of the natural world and as the expression of the interconnectednes of all forms of life, recalling earlier studies by Uexküll and by Deleuze and Guattari.[53] Indeed, the chapter titled 'Crève-Cœur' opens with the statement: 'Sous le mangier les notes volent. À la rivière d'Alma, du côté de la ravine, je les entends' (p. 75). Flight here may suggest a relation between tree and water, earth and sky. It might give expression to Merleau-Ponty's statement 'Umwelt, c'est une mélodie qui se chante elle-même.'[54]

Le Clézio then goes on to present us with details of Dodo's background as a musician: his Creole mother, Rani Laros, was 'une grande chanteuse' (p. 151), and as a child he played the piano. Later, in the absence of an instrument, 'je garde les notes dans ma tête et quand je veux, je les fais voler' (p. 76). These images of words or sounds flying through the air are reminiscent of Vinteul's septet[55] in Marcel Proust's novel, a writer with whom Le Clézio has declared close affinities and whose direct influence can be detected in *Alma*.

Indeed, music becomes the trigger for an intricate web of relations and is perhaps the most comprehensive rendering of Deleuze's concept of 'becoming' in the book. In the first instance, we learn that a 'mongol' girl, one with little understanding of words, comes to listen to Dodo as he sings the notes of Felix Mendelssohn's 'Romances sans paroles'. Here, it

52 See, for example, Édouard Glissant, *Le Discours antillais* (Paris: Gallimard, 1997), p. 407. The Creole language is in essence musical.

53 von Uexküll, *Milieu animal et milieu humain*, pp. 108–113 and Deleuze and Guattari, *Mille Plateaux*, pp. 381–433. See also Chapter 1, p. 8.

54 Merleau-Ponty, *La Nature*, p. 28.

55 See Marcel Proust, *À la Recherche du Temps Perdu* (Paris: Gallimard, 1954), pp. 347–348. The text is referring to 'la petite phrase'.

is the individual notes of the music that convey an intense sensorial, synesthetic experience. At one point, the girl's body becomes transfigured: it is pure sound:

> Sous le manguier, la petite fille mongolienne vient écouter, les notes l'attirent en volant dans l'air, les notes sont de toutes les couleurs, elles ont le goût des bonbons et du miel, mais aussi par moments le goût de la pluie et du vent des cyclones, pour elle je chantonne la musique, même si elle ne comprend pas je crois qu'elle les sent, pas avec des mots, avec des sons, juste à l'arrière de la bouche, les dents serrées, hm-hm, lan lan, hm-hm- les 'Romances sans paroles' de Mendelssohn, ça c'est bien puisqu'elle ne peut pas comprendre les paroles. (p. 76)

Images of flight and sound are repeated a few lines later and conveyed in the rhythm of the sentence 'je peux seulement faire voler les notes, dans la ravine, sous le manguier, pour elle'. At the same time, references to water, air and wind in this passage would clearly suggest an identification of music with the cosmic and with a primordial energy. Interestingly, in the chapter headed 'Dodo', we learn that the very name 'Alma' derives from 'Araceli', meaning 'la musique du ciel' (p. 130).

Music, however, is not only presented as a sensorial experience. It is also evoked in terms of cultural interaction, that is, of a Relation between different nationalities. This is conveyed in the use of proper names, those of composers, singers and songs. Indeed, as earlier, naming has an expressive, poetic function. A central role is played here by the song 'Auld Lang Syne' to which there are multiple references in the text. Indeed, its function is that of a ritornello[56] or refrain. We also learn that the Creole narrator sings it in Scottish to Céminor, that the music is by the German Franz Schubert, and that Dodo used to play and sing it for his Scottish grandmother, Beth! (p. 76). These multiple names are then repeated like a litany when, on New Year's Eve, Dodo enters a church where he plays and sings 'Auld Lang Syne' to the pianist, Michel. However, at this point, the reader is provided with additional information, a variation on the tune: we now learn that the words are by Robert Burns (pp. 78–79). Once again, the

56 See Deleuze and Guattari, *Mille Plateaux*, for a study of the ritornello (pp. 385–391). For a discussion of this concept, see Adina Balint, *Le processus de création dans l'œuvre de J. M. G. Le Clézio* (Leiden: Brill, 2016), pp. 133–134.

effect is that of a chant, a striking contrast with what will later emerge as a very bleak presentation of contemporary Western societies, imprisoned within the art/life duality.

Furthermore, in this same episode, song, like touch before, becomes a mode of communication with the dead and is indissociable from memory: it was immediately after visiting the grave of his parents on New Year's Eve and writing over their names that Dodo entered the church and sang 'Auld Lang Syne', the very words of the song invoking the dead (p. 77). It would suggest the power of music to collapse the boundaries of time and space.

The magical power of the sound of individual words is also explored later in a meeting of Dodo and the tramps of Paris in a building at Saint-Germain-en-Laye (p. 203). After the official, rather impersonal welcome by the priest, Père Antoine,[57] authority is handed over to a young man who reads aloud a poem[58] of the Spanish poet Ruben (p. 205).[59] For Dodo, the sounds of the words in Spanish awaken buried memories of notes of music, 'les sons de l'autre côté du monde', those of the material ground of life (p. 206). When he is invited to play the piano and sing, he chooses 'Auld Lang Syne': in an explosion of sounds, he collapses the opposition between life and death, past and present. In sympathy with the song, the tramps themselves join in the singing:

> Je joue mon morceau, je suis capable de jouer, le vieux *Auld Lang Syne,* de tout mon cœur, mes mains tordues caressent le clavier blanc, et la musique sort de mes doigts et remplit la salle, je joue pour vous dire adieu, je ne vais plus vous voir, adieu, adieu, c'est dans la chanson de Schubert, adieu à l'amour, et les clochards commencent à chanter avec la musique, ils battent les mains, ils crient, je ne sais pas si ce sont des 'hourras' ou des 'hou!', je joue, et quand j'ai fini de jouer, je descends de l'estrade, je traverse la salle, avant que Père Antoine me parle, je m'en vais. (p. 207).

57 See pp. 115 and 112 of this chapter.

58 As Césaire has written:'La poésie révèle l'homme à lui-même. Ce qui est au plus profond de moi-même se trouve certainement dans la poésie [...]. C'est le poème qui me le révèle et même l'image poétique'. See Aimé Césaire, *Nègre je suis et Nègre je resterai* (Paris: Albin Michel, 2005), p. 47.

59 Rubén Darío (1867–1916) was a Nicaraguan poet who initiated the Spanish-American literary movement known as modernism.

This is an expression of pure energy and would suggest the healing, transcendent power of song. Dodo senses at this point that he is approaching the end of his journey, the image of the sea suggesting a return to the source of life, the space of origins. He stops playing and leaves:

> Je m'en vais loin, maintenant, dans les rues et sur les routes, je marche sur la route de Palma qui descend vers Flic en Flac jusqu'à la mer, je marche jusqu'au bout de la route. Je marche vers la fin de mon voyage. (p. 208).

The four main clauses with the repetition of 'je marche' convey the rhythm of his movements; they also evoke the sound of a drum, the expression of pure energy.[60] Indeed, it is the reiteration of the same sounds that becomes a striking feature of the first two sections of this study, those devoted principally to Creole culture and to interculturality.

The Sociopolitical

Importantly, the Creole ecological vision, one of Relation with the natural world and with different cultures, is presented as essentially fragile as it comes into conflict with the forces of violence and social exclusion both in Mauritius and later in Paris and Nice. Le Clézio could be said here to strengthen his own poetic liberating vision of the world by contrasting it with the dire consequences of (a) Western postcolonial expansion and globalisation, and (b) Western economic and political thought rooted in the supremacy of the human and in the body/mind, nature/culture divide.[61]

60 Glissant compares the Creole language to the rhythm of the drum. See Glissant, *Le Discours antillais*, pp. 406–407. It is the intensity of sound that dictates meaning.

61 See Graham Huggan and Helen Tiffin, *Postcolonial Ecocriticism: Literature, Animals, Environment* (New York: Routledge, 2015). The authors highlight the disastrous impact of global capitalism, anthropocentrism and racism.

Ecological and cultural violence: Mauritius

Indeed, in his travels around Mauritius, Dodo encounters a very real threat to his bird identity, to his relationship to the land, to the very earth. This hostility manifests itself in a destruction of Creole spaces, in an act of dispossession that also characterised the era of slavery and led to the original death of the birds who could walk on the island. For instance, in the chapter called 'Alma', we learn that the little hut on the Alma estate where the Dodo lived as a child with his Creole mother and Creole nurse, Artémisia, was destroyed by French property developers – les Armando – and reduced to ashes (p. 68). The later reliving of this act conveys the traumatic impact of this loss of roots, one that is rendered more immediate by the use of the present tense throughout. Indeed, we learn that before leaving for France, Dodo returns to Alma: 'je cherche l'endroit où Artémisia a sa case, où elle raconte ses histoires et ses sirandanes, ensuite le bulldozer des Armando écrase tout, et elle quitte le monde pour aller au paradis' (p. 166).[62]

Furthermore, this eradication of cultural memory is also displayed in attitudes to the dead. The Creole Yaya, the daughter of a slave, has no right to a grave or a tombstone: 'Elle n'est pas à Saint-Jean. Elle n'est nulle part. Elle n'existe pas' (p. 63). Indeed, in Creole cultures, it is the name itself that becomes not only the guardian of memory but also a vehicule for the resurrection of life, a collapsing of the life/death dichotomy on which classical Western thought is based. Significantly, before Dodo leaves for France, he visits the Saint-Jean cemetery and rewrites the names of his parents over their graves, aware that 'après moi plus personne ne va écrire, la pluie et le vent effacent les noms et les dates et ils n'habitent plus jamais la terre' (p. 167).

This cultural violence is sharpened in a vivid presentation of the class-ridden capitalist society of contemporary postcolonial Mauritius where economic prosperity is based on social inequality, a theme explored in depth

62 This destruction of Creole culture is evidenced in Jérémie Felsen's account. Here we learn of the destruction of the land, the death of plants and pollution of the sea. See pp. 94–95.

by Jérémie Felsen in the interwoven narrative. In 'Le Grand Cimetière de l'Ouest', Dodo is attacked by a group of students from the wealthier district of Mauritius, who, after mocking his name, throw rocks at him and laugh at his face, 'Pe'sonne! Vilain couma macaque, enn guèle macaque!' They also beat him with a cricket bat, an attack which he was lucky to survive (pp. 130–131).[63] Indeed, as a victim of extreme violence by the privileged middle classes, Dodo is in danger of suffering a fate similar to that of the original dodos on the island who were slaughtered by the Western colonial forces.[64]

Ecological and cultural violence: Paris

Encouraged by Vicky, who has contacts in the capital, Dodo is persuaded to fly to Paris: 'Dodo, tu vas connaître beaucoup de nouvelles choses, tu vas rencontrer beaucoup de gens nouveaux' (p. 164). However, upon arrival in the city, Dodo encounters a greater degree of cultural violence, a sustained threat to his bird identity, that is, to the Creole ecological vision as experienced in the early stages of his journey around Mauritius. Any fusional sensorial relationship with the earth or with the cosmic is now completely out of the question: the equilibrium between life and death has indeed been destroyed. In contrast to the rain in Mauritius, the rain in Paris has no taste or smell: 'Je sais que je voyage très loin, parce que je n'ai pas le goût de l'eau, je ne sens pas l'odeur de l'eau, et ça me fait un pincement au côté parce que j'imagine l'odeur de l'eau dans l'île' (p. 181).

When he goes down to the river, he becomes the immediate target of sensorial aggression evoked initially in images of extreme cold:

63 This violence has contemporary resonances. A Creole singer was murdered in police custody in 2012. See also the chapter called 'Pomponette' of the interwoven narrative where Jérémie Felsen presents a powerful critique of racism on the island. See pp. 142–148.

64 Jérémie Felsen finds in his father's collection the stone of the gizzard of the original birds that could walk, suggesting his own family complicity in their slaughter and indeed in the slave trade. See p. 35.

> Je prends l'escalier en pierre qui va à la rivière, le vent froid met des larmes dans mes yeux, les larmes coulent sur mes joues, sur le bout de ma langue. Mes mains sont froides aussi, je les mets dans les poches de l'imperméable. (p. 181)

The physical impact is heightened here in images of touch and in the close attention accorded physical detail: eyes, cheeks, tongue, hands. Furthermore, instead of being a source of life, the water has become an agent of destruction:

> Le fleuve fait un bruit que je ne connais pas, un bruit léger, je vois les tourbillons qui entraînent les feuilles mortes, les branches cassées, et aussi des détritus jaunes, je vois même un animal mort, un chien noyé, son ventre gonflé, ses pattes raidies, il s'en va en roulant, il disparaît. (p. 181)

Here the use of juxtaposition and the accumulation of detail conveys the enormity of the violence. The sensorial impact is again strengthened in the reiteration of the same short, sharp sounds ('cassées', 'noyé', 'gonflé'). In addition, pollution is now expressed in terms of smell:

> Je prends l'eau dans mes mains, je sens son odeur, mais ici ce n'est pas la même odeur, c'est une odeur de cendre, une odeur de pisse, une odeur de mort, mais ce n'est pas l'odeur du cimetière Saint-Jean, peut-être l'odeur du cimetière de l'Ouest, une odeur lourde, la pisse du monde, d'une ville, d'un pays. (p. 181).

Here the intensity of the corruption is conveyed not only in the lexical repetition and choice of images but also in the rhythm of the sentence, in a progression from the immediate towards spatial expanse, 'la pisse du monde, d'une ville, d'un pays'. Dodo is indeed in danger of losing his bird identity and becoming a 'one-dimemsional man'.[65]

As in a musical score, Le Clézio returns to this evocation of Paris three chapters later in 'Le prophète', in a process of expansion and intensification:

> La ville est mon île maintenant, que ne borde pas sur la mer, mais les autoroutes qui ronflent et grognent avec le bruit des vagues sur les récifs, les falaises blanches des immeubles de douze étages, aux mille fenêtres, les terrains vagues et les talus du

65 See Herbert Marcuse, *One-Dimensional Man: Studies in the Ideology of Advanced Industrial Society*, ed. Jacob T. Levy (Oxford: Oxford University Press, 2017).

chemin de fer, les ponts noircis par la suie, les forêts hérissées où s'accroche les sacs en plastique. (pp. 252–253)

Sensorial aggression is now experienced in visual terms of height, quantity and size as well as in those of sound, that of the motorways 'qui ronflent et grognent'. The question of pollution is again addressed in the image of plastic bags clinging to the trees, clearly a threat to the Creole concept of the sacred as presented in the early stages of the narrator's journey.

These renderings of the contemporary city can be interpreted as a sustained critique of capitalism, notations of size and quantity resonating very closely with passages from Le Clézio's earlier book *Les Géants*.[66] Indeed, there is a similar focus on cars ('les moteurs') and on the image of the machine, a word that also denotes people (p. 224), suggesting a total eradication of the human. In addition, both novels draw our attention to pollution and to the destruction of the very fount of life.[67] We may be reminded here of the writings of the political ecologist André Gorz,[68] for whom the impulse to maximise production and economic growth poses a very real threat to the ecological balance, a theme explored earlier in *Haï* where Le Clézio warns us of an apocalypse (pp. 107–109). As Bruno Latour has written, the era of the Anthropocene has already arrived, humans are changing the function of the earth's biological systems.[69]

This environmental destruction[70] would clearly have its roots in contemporary Western thought, in the promotion of a hyper-rationality and of the technological, the outcome being what David Abram has described as a general anaesthesia, a distancing of the self from the activity of the senses,

66 See J. M. G. Le Clézio, *Les Géants* (Paris: Éditions Gallimard, 1973), pp. 224–228.

67 Ibid. p. 231.

68 André Gorz, *Capitalism, Socialism, Ecology* (Brooklyn, NY: Verso Books, 1991) and *Écologie et Politique* (Paris: Éditions Galilée, 1977). For a discussion of his thought, see Louise Hardwick, 'Towards Biopolitical Ecocriticism: The Example of the *Manifeste pour les 'produits' de haute nécessité*', *French Studies*, Vol. LXX, no. 3, pp. 375–377.

69 Latour, *Face à Gaïa*, pp. 147–191.

70 As Jérémie Felsen explains, environmental destruction has also become a feature of contemporary Mauritius: the noise of a jumbo jet is compared to that of the cannon fire and death of the dodos during the colonial era. See p. 41.

from feeling.[71] As Christopher Johnson has also explained, this privileging of the machine would make us less not more human, it would signal a loss of empathy and weakening of the imagination, the trigger for violence, both towards human and non-human forms of life.[72] As we saw in *Haï*, it is only through the activity of the imagination, through the expression of a primordial energy, that humans can communicate with the non-human in a 'becoming-animal'.

Indeed, as Dodo has discovered in Paris, the contemporary city poses a threat to his Creole bird identity and to his experience of the sacred, as conveyed in his relationship to the cosmic forces and to the earth. Parallels can be established with Le Clézio's earlier book *Révolutions* where he associates contemporary Nice with the death of *le vivant*, with 'des palmiers immobiles dans leur pots' (p. 19), and with the swallows giving out 'des sifflements inquiétants, oppressants [...] dans le silence engourdi du dimanche après-midi' (p. 89). And, as in Paris, the water itself is polluted: the sea becomes 'une mer épaisse, lourde' (p. 94). Le Clézio is clearly drawing attention in his fiction to the human responsibility for ecological destruction: it is a call for action, for transformation of the self, nothing is inevitable.

Institutional hostility: Paris

Upon his arrival in Paris, Dodo experiences a much greater degree of institutional hostility and of social alienation, the outcome of Western values and of the prestige accorded property and possessions. For the tramps of Paris, there can be no question of 'becoming-bird'. They are presented as passive victims of authority, as innately inferior to the rest of the population and, like children, unable to look after themselves (pp. 182–183).

71 David Abram, *Becoming Animal: An Earthly Cosmology* (New York: Vintage Books, 2010). See, for example, pp. 67–72.

72 Christopher Johnson, *System and Writing in the Philosophy of Jacques Derrida* (Cambridge: Cambridge Studies in French, 1993). For a critique of the machine, see also Le Clézio's short story 'A peu près apologue', in *Histoire du pied et autres fantaisies*, pp. 337–338.

Indeed, this arrogance displayed towards the homeless is illustrated in the first instance in the figure of Père Antoine in a meeting he conducts with the tramps of Paris. Here, his approach contrasts sharply with the kindness displayed by the staff of the Swedenborg church and the love communicated in the ceremony conducted by Père Chausson. The chapter titled 'Dodo voyage' opens with the sentence: 'Père Antoine commande la rencontre avec les clochards de Paris' (p. 203). The very pronouncing of orders here would suggest the authoritarian, the absolute power accorded the Church as guardian of the truth and of morality. Indeed, the tramps are presented as social outcasts, as if they were contaminated. The building to which they have been summoned is at a considerable distance from the centre of the city: 'C'est dans une grande salle, loin de tout à Saint-Germain-en-Laye, c'est le nom qu'on me dit dans le train' (p. 203). In addition, the chairs and tables are arranged as in a military establishment and the tramps have to enter the room two by two. Like a commanding officer in the army carrying out a role call, Père Antoine reads out their names, however only their forenames because 'personne ne doit connaître leurs noms de famille ni d'où ils sont' (p. 203). Instead of empowering them, of instilling a sense of worth, he is drawing attention to their divorce from the wider community as if they were pariahs. He also resorts to abstract generalisations, to clichés that might 'sound good' but only serve to reinforce their identity as tramps living on the edges of society: on hearing their name, each person must stand up, make a hand signal and smile, 'parce que nous sommes tous frères et sœurs, c'est la grande famille des SDF, des clochards sans frontières' (p. 204). As Dodo comments, once they have eaten, very few are likely to meet again, 'parce que les rues sont sans fin, la ville est sans fin, ils marchent tout le temps, puis ils s'asseyent par terre là où ils sont, ensuite ils se relèvent et ils marchent encore' (p. 205).

Indeed, unlike Père Chausson or the members of the Swedenborg church, Père Antoine can offer no real friendship or significant help. The encounter has not been in any sense transformative. At one point, with Dodo in mind and resorting once again to generalisations, the priest evokes the enormous gap between rich and poor on the island of Mauritius, unaware that the problem could be on his very doorstep. Indeed, he is moved by his own rhetoric: 'Quand il a fini de parler, Père Antoine se mouche parce qu'il est ému, il essuie les verres de ses grosses lunettes' (p. 206). Moreover, there is no recognition on his part of the Creole roots of many of these tramps, no question of Relation. As we learn later, there were Arabs amongst the tramps, one of which – Béchir – will later befriend Dodo. Le Clézio is mounting a very strong critique here of the claim to a single absolute truth made by any religious institution, denouncing, at the same time, the stance of superiority adopted by its leaders. He is also condemning widespread attitudes towards the poor and homeless in Western capitalist societies, in particular the belief in the inevitability of poverty: Père Antoine sends the tramps back onto the streets. As both Thomas Piketty and Naomi Klein have argued, Western neoliberal economics is rooted in inequality.[73]

The streets of Paris and Nice as the expression of Western values

Tramps

In stark contrast to Dodo and to 'la fille aux cheveux bleus', the tramps in Paris lead a rather sedentary life; they are not on any journey: 'Eux, les clochards, les SDF ils ne savent pas marcher. Ils arrivent quelque part et ils n'en partent plus' (p. 248). They even construct their homes out of canvas and pieces of wood, a reflection of the importance attached to property ownership and to a single permanent residence in the West: Le Clézio writes that 'À Paris, les dodos ont beaucoup de bagages' (p. 253).

What is more, the best spaces in Paris are reserved 'pour les clochards d'ici': if an outsider approaches, 'ils viennent te menacer, te battre parce qu'ils croient que ces endroits sont à eux' (p. 184). There is no question of sharing: personal interests take precedence over 'l'intêret commun', an attitude that Badiou would describe as a capitalist view of the human, one

73 Thomas Piketty, *Capital in the Twenty-First Century* (Cambridge, MA: Harward University Press, 2014), trans. Arthur Goldhammer. See also Naomi Klein, *The Shock Doctrine: The Rise of Disaster Capitalism* (London: Penguin, 2007). Klein also argues that free-market fundamentalism relies on human disaster and human suffering to function. She cites the example of the millions of children in the United States who go to bed hungry.

that by its very nature ignores any question of justice or equality.[74] This violence, or threat of violence could also be interpreted as a *mise-en-abyme* of nationalism, or more pertinently, as a declaration of war.

However, at the same time, and unlike Dodo, who is constantly on the move, the tramps themselves become the easy target of hostility from the 'forces of law and order'. It could be the aggressive security guards who 'rôdent avec les chiens méchants' demanding their names (p. 248).[75] It could also be the police, who, although more polite, are potentially more dangerous: 'Bonsoir, contrôle d'identité, vos papiers. Vous êtes français ? Oui ? Avez-vous une pièce d'identité, s'il vous plait ?' (pp. 248–249). These scenes would suggest a direct critique of the concept of a single identity, or a single allegiance. They also illustrate more generally the relationship between nationalism and social exclusion.

Furthermore, the attitude and behaviour of the French authorities here fosters prejudice, creating a negative public image both of the homeless and of the 'foreigner', one that could by its very nature trigger violence. This is illustrated in the final scene of this section where the tramps become the target of a hunt, all the more ironic in a city that prides itself on being the beacon of civilisation. Here the intensity of the violence and the sense of mounting danger are conveyed in a seven-line sentence:

> Ils (les loubards) circulent la nuit, en bandes de cafards, ils sont en voiture, à moto, quelquefois à pied, alors les dodos doivent se cacher, ils restent collés ensemble en bas des immeubles, ou sous les ponts des autoroutes, là où beaucoup de gens passent, ils s'enveloppent dans leurs plastiques pour disparaître, ils font des tas de cartons et de bouts de caisse, ils croient qu'ils sont invisibles. (p. 251)

Scenes as above could be said to historically contextualise Glissant's concept of Relation, highlighting its contemporary relevance. Indeed, Le Clézio will go on to develop this critique of spatial exclusion and of social and political alienation.

74 See Alain Badiou with Aude Lancelin, *Éloge de la politique* (Paris: Flammarion, Café Voltaire, 2017), p. 41.

75 Far from being an expression of Relation, as earlier, naming becomes an agent of oppression.

Child refugees and the victims of war

In the same chapter, Le Clézio presents us with the image of homeless child refugees and gypsies roaming the streets of Paris. Indeed, everywhere Dodo goes – and like the Pied Piper – he is followed by children, hoping no doubt that he is leading them somewhere. It is his bird-like qualities that attract them; they address him by his name: 'Dodo! Dodoo! Ils chantonnent mon nom, ils courent, ils répètent: Dodo! Doo-o-oh!' (p. 256).[76] As asylum seekers, they have lost their roots, their relationship to land, culture and family.[77] Le Clézio writes:

> Ils ne vont nulle part, leur maison est nulle part. Les Roumains, les Yougos, les Gitans, les Arabes, les Sénégalais, les Afghans. Ils sont chassés de tous les pays, ils n'ont pas de famille. Ils vont en Angleterre, en Allemagne. Ils ne savent pas où. (p. 256)

Many of these names refer to countries or to populations with which the West has been at war, and/or colonised, and the reference to children may point to their status as orphans. Their actual presence on the streets would bear witness to the rising wave of racism and nationalism in contemporary Western capitalist societies, and to a widespread refusal to offer hospitality. As Dodo sadly comments, 'ils imaginent que je les conduis quelque part' (p. 256). In a nine-line flowing sentence, Le Clézio compares their movements first to a river and then to a flood. Intensification as well as a sense of inevitability is conveyed through the listing device and in the very rhythm of the sentence cumulating in a climax: indeed, the text calls to mind the miners' strike in Émile Zola's *Germinal*,[78] where

76 This quotation from the chapter titled 'Le prophète' has biblical connotations. The image here of Dodo being followed by the poor, by the outcasts, and later his act of curing a dying child through touch, may call to mind the figure of Christ: 'je passe mes mains sur le visage du bébé, je souffle dans les trous de son nez et l'enfant se met à tousser' (p. 260).

77 Referring to the French attitude to migrants, Le Clézio writes: 'La politique est un monstre froid; elle agit en suivant des lois et des instructions qui ne tiennent pas du sentiment humain'. See 'Un Déni D'Humanité Insupportable', in 'LOBS/N 2775-11/01/2018.

78 Émile Zola, *Germinal*, trans. Roger Pearson (Penguin Books, 2004).

the march of the minors through the streets of Paris is likewise presented as a flood, 'the roaring torrent of strikers' (p. 334) and as an outpouring of energy (p. 339). Le Clézio writes:

> La foule marche avec moi, tous ces foucas, ces va-nu-pieds, ces clochards, ces enfants voleurs, alors les gens s'écartent pour nous laisser passer, le fleuve brun doit couler, l'eau sale doit suivre les ruisseaux, personne ne peut l'empêcher, personne ne peut l'ignorer, il faut que ces anoraks, ces jeans, ces vestons, ces bonnets de laine, ces cagoules, ces chaussures éculées, il faut qu'ils passent, la vanne est ouverte et l'eau doit couler sur le trottoir, suivre les rigoles et les fissures. (p. 257)

War can be viewed here as the outcome of Western neocolonialism and of its hierarchical ethnocentric concept of history. At the same time, and in contrast to the poetic language of *Créolité*, it could be seen as the product of 'l'image dogmatique de la pensée', one rooted in presuppositions, in abstractions and in an inability to conceive of difference.[79] In other words, in the West political power is engrained within language itself.[80] In 2012, Le Clézio created with Issa Asgarally, the Fondation pour l'interculturel ou la paix. Here, in a powerful condemnation of Samuel Huntington's book *The Clash of Civilizations*,[81] they stress the role of literature, of the imagination in overcoming cultural barriers. Le Clézio himself has also been involved in direct political action, signing petitions and publishing articles pleading the cause of migrants and of refugees.[82] At the same time, in his book *Frères migrants*, Patrick Chamoiseau refers to 'la barbarie

79 See Gilles Deleuze, *Différence et Répétition* (Paris: Presses Universitaires de France, 1968), p. 192.

80 Le Clézio has stated that European philosophers have done nothing to prevent war and racism. He does, however, praise thinkers like Badiou and earlier, Jean-Paul Sartre, who seek to express philosophy in everyday life. See Lu Zhang, 'Entretien avec J.M.G. Le Clézio', *Les Cahiers J.-M. G. Le Clézio* no 8, p. 99.

81 Samuel Huntington, *The Clash of Civilizations* (New York: Simon and Schuster, 1996).

82 See 'Un Déni d'Humanité Insupportable'. L'OBS/N2775-11/01/2018. Here Le Clézio quotes Martin Luther King Jr: 'Nous avons appris à voler comme des oiseaux et à nager comme des poissons, mais nous n'avons pas appris l'art tout simple de vivre ensemble comme des frères'.

nouvelle' (p. 31) and 'la barbarie des frontières' (bookflap), concluding with a 'Déclaration Des Poètes'.[83]

Furthermore, and in stark contrast to Dodo's poetic vision, war is the outcome of a mode of thinking that does not recognise the intrinsic value of life itself. Its very horror, that is, the dire human consequences are illustrated in the story of the Arab Béchir whom Dodo met in Paris.

The impact of war on the individual

Indeed, in the figure of the Creole-speaking Béchir, a refugee from the Algerian war of independence, Le Clézio is bringing to the fore the traumatic impact of Western colonial wars. Suffering from a fatal disease and with a bullet lodged in his brain, Béchir is hoping to go home to die: 'Chez moi, c'est à Tlemcen, en Algérie, de l'autre côté des montagnes c'est Oujda le Maroc. C'est là que je dois mourir' (p. 294).[84]

The episode opens by drawing attention to Béchir's reasons for following Dodo: it was the beauty of his playing of 'Auld Lang Syne' at Saint-Germain-en-Laye in Paris, that prompted Béchir's decision to accompany the narrator and to recover a lost identity: 'je t'écoute et je pleure, je sais que je vais partir avec toi, je vais prendre le bateau pour retourner à Tlemcen, dans ma famille' (p. 294) and 'je sais que c'est maintenant, je dois marcher avec toi jusqu'à la mer, je veux trouver l'endroit de ma mort' (p. 295). In this very reference to 'Auld Lang Syne' and to the sea, Le Clézio is making an association between song and memory, between song and a return to the space of origins, the fount of life. The text would recall Césaire's *Cahier d'un retour au pays natal* where sound/dance plays a similar function.[85]

Indeed, from the very outset, Le Clézio presents us with a number of details about Béchir's past and, significantly, about his state of mind. In

83 See Patrick Chamoiseau, *Frères migrants* (Paris: Éditions du Seuil, 2017). Chamoiseau and Le Clézio have collaborated in a number of enterprises.

84 We now learn that the sun and rain have effaced the white bird image on Dodo's bag (p. 294). Like Béchir, he is moving towards the end of his own journey.

85 Aimé Césaire, *Cahier d'un retour au pays natal* (Tarset: Bloodaxe Books, 1995), p. 132. Le Clézio quotes from this book in his novel *Poisson d'or* (Paris: Gallimard, 1997), pp. 146–147. The words are also sung by the protagonist, Simone from Haiti.

the first instance, we learn that he is an Algerian, the son of a harki who fought in the Algerian war (p. 253) and that later he left with his father for 'le camp des harkis au sud de la France' (p. 295) where his father died.[86] An unwillingness to speak on his journey from Paris to Nice, together with the fear of being alone would suggest that he is suffering from acute trauma: indeed, the text might bring to mind some of the horrific scenes described by Le Clézio in his earlier book *Révolutions*.[87] Béchir is clearly the victim of what Césaire has described as the European descent into barbarism in his *Discours sur le colonialisme*.[88] The repetition of 'ne … pas' in the passage quoted below is an affirmation of his traumatised state:

> Béchir aime marcher avec moi, parce que je ne parle pas, je ne raconte pas ma vie, je ne lui pose pas de questions sur sa vie, ça ne me regarde pas. La nuit, je ne dors pas, je reste assis les yeux ouverts pendant que Béchir ronfle, ça le rassure, je suis son chien de garde. (p. 254)

Béchir would seem to be completely adrift in the world, lacking any relationship, either with people or the landscape: 'il cache sa tête dans un sac de papier avec des trous pour respirer, il enfonce son bonnet de laine jusqu'aux yeux même s'il fait chaud, pour ne pas voir le ciel et les étoiles au-dessus de nous' (p. 296). One can only imagine what his past might have been.

Significantly, it is Le Clézio's close attention to physical detail in this episode that heightens the sensorial impact, conveying the magnitude of Béchir's suffering. One morning, Dodo finds Béchir lying on the ground,

86 Harki is the name for native Algerian Muslims who remained loyal to the French during the Algerian war, 1954–1962.

87 In his book *Révolutions*, Le Clézio describes the French firebombing of an entire village in Algeria (pp. 206–207), as well as the death of a prisoner impaled on a fence and eventually shot (pp. 210–211). He likens the fate of the Arabs to the public punishments inflicted on the maroons in Port Louis and to the beating to death of slaves (pp. 85–87). The text suggests that Béchir himself may have been witness to acts of torture.

88 'Passant plus outre, je ne fais point mystère de penser qu'à l'heure actuelle, la barbarie de l'Europe occidentale est incroyablement haute'. See Aimé Césaire: *Discours sur le colonialisme* (Paris: Présence Africaine, 1955 and 2004), p. 29.

lifeless: 'Lui, il reste par terre dans le champ de blé, ses mains sont froides et ses lèvres sont bleues' (p. 296). Afraid that he is dead, Dodo shouts at him and then resorts to the power of touch, to the communication of energy, the lifegiving. The rhythm of the passage mirrors that of Béchir's awakening:

> Enfin Béchir ouvre les yeux, ses yeux sont troubles, d'un vert sale, les paupières collées par les larmes. Il répond quelque chose que je ne comprends pas dans sa langue du bled. Le soleil le réchauffe, moi je le frictionne sur les jambes, sur la poitrine. (pp. 296–297)

Le Clézio's presentation of Béchir would indeed support Glissant's view that modern warfare has lapsed into 'unspeakable horrors', citing as an example Algeria.[89] It would also illustrate the power of literature to communicate this horror and hopefully, in the words of Arthur Rimbaud, to 'changer la vie'.[90]

Nice and racism

Sadly, as it emerges, Béchir becomes the fatal victim of a violent racist attack at the station in Nice.[91] The city has become, as it were, the end of the journey for both of them. As Dodo declares: 'C'est la fin du voyage. Je n'ai plus besoin de marcher, jamais' (p. 301), adding:

> Béchir lui non plus ne voyage pas. Là-bas, de l'autre côté, dans son bled à la frontière de Tlemcen, sa famille l'attend mais il est mort. Il est mort sur le port, sans rien dire, couché sur son carton, avec son bonnet de laine sur les yeux, et son sac en papier troué pour respirer, mais il ne respire pas. (p. 302)

Here, the use of repetition of words and sounds ('mort', 'port') ('troué', 'couché'), as well as the rhythm of the final sentence with its abrupt ending, conveys the intensity of the tragedy, of the act of loss. It would indeed evoke a litany. The passage is followed by a highly detailed yet

89 Édouard Glissant, *Introduction to a Poetics of Diversity* (Liverpool: Liverpool University Press, 2020), trans. Celia Britton, pp. 70–71. Glissant adds that 'we will need to invent new ways of resisting', ibid.

90 See Arthur Rimbaud, *Oœvres* (Paris: Éditions Garnier Frères, 1960), p. 225.

91 A parallel can be drawn between Nice and Paris. Both are hotbeds of capitalism associated with wealth and tourism as well as with extreme poverty.

moving description of his dead body where the traditional boundaries between biological realism and the poetic collapse. The focus on colour and on temperature strengthens the emotional impact:

> Il est mort comme mon papa, avec la peau de son visage qui est toute blanche, ses yeux ouverts qui ne voient plus, sa bouche sèche et noire, et le froid dans ses mains, dans ses jambes, même les polis gris de sa barbe sont immobiles. (pp. 302–303)

This rejection of cultural difference as well as disregard for the poor and homeless is further illustrated in the refusal of the French authorities to send Béchir's body back to Algeria in accordance with the Arab tradition. Instead, it will be sent to a communal grave: 'On met le corps dans un cercueil en planches, et puis on coule de la chaux vive. On n'écrit pas de nom sur une pierre ni rien du tout' (pp. 302–303). Dodo adds, 'Pour moi aussi c'est pareil' (p. 303). There will be no recognition of his own Creole origins, no ritual of memory or communication with the dead as witnessed in his earlier visits to the cemeteries in Mauritius and to the tree under which the Creole Yaya is buried.

Indeed, the death of Béchir has led Dodo to the following political conclusions. The passage contains echoes of a speech of Vladimir Lenin:[92]

> Alors à ce moment-là, ici et là-bas, c'est la même chose. Je ne le sais pas avant de voyager jusqu'en France. Les gens croient qu'ailleurs est différent. Mais ailleurs c'est pareil, il y a des grands et des petits, il y a des importants, les présidents, les directeurs, les banquiers, les Armando, les Escalier (...) tout bann zan-là. Et il y a les autres qui ne valent pas une roupie, les oubliés, les écrasés, ils n'ont pas de carte de visite, ils n'ont pas de carte de crédit, ils n'ont rien dans les poches, juste quelques billets en lambeaux et quelques sous rouillés. Et maintenant je le sais, parce que quand Béchir est fin mort, il reste sur le brancard dans le couloir de l'hôpital, et les toubibs en blouse blanche et les infirmiers en blouse verte passent devant lui sans le regarder, alors moi je pars sans rien dire et je marche dans la nuit, et dans le sac à dos de Béchir je ne trouve rien qu'une carte de soldat et son gros bouquin vert. (pp. 308–309)

92 See J. M. G. Le Clézio, *Poisson d'or* (Paris: Gallimard, 1997), p. 207. Referring to the content of her baccalaureate, the protagonist, Laïla, quotes a few lines from a speech of Lenin.

Le Clézio is calling into question here the fundamental premises of Western humanism as expressed in the French motto 'liberty, equality and fraternity'. He is not only drawing attention to the enormous gap between rich and poor, a feature of contemporary Western capitalist societies, but also, importantly, to a widespread social indifference. In his earlier focus on the Creole poetic vision and on the very notion of Relation, Le Clézio is positing the political power of the imagination, the need to create new social structures, new ways of being in the world. As Deleuze has also stated, 'penser, c'est créer, il n'y a pas d'autre création'.[93] And for Badiou, our only hope lies in a revolutionary becoming, a creation of the new, 'création immortelle d'un événement'.[94]

Paradise or Time Regained

The concluding chapters of *The Story of Dodo* take the reader back in circular fashion to the Creole culture of Mauritius and to a celebration of life and beauty, of the healing, transcendent powers of the imagination, a reaffirmation of Dodo's bird identity. He has not allowed himself to become the prisoner of history, of the pain or horror of war. I shall begin by developing the theme of art embodied earlier by 'la fille aux cheveux bleus' and will conclude with a return to that of music, a major structuring device of the narrative whose tempo is now accelerating. The final section highlights the need for constant repetition, for the reassertion of *le vivant* and of the cycle of life.

93 Gilles Deleuze, *Différence et répétition* (Paris: Presses Universitaires de France, 1968), p. 192.

94 Alain Badiou, *L'éthique* (Caen: NOUS, 2003), p. 12.

Art

In the chapter headed 'La mer', we learn of Dodo's meeting at the port of Nice with an artist whom he immediately recognises as a Fe'sen, that is, as belonging to the French side of his family, those responsible for his exile from the property, Alma (pp. 305–311). However, when Dodo gives his name, 'Moi c'est Fe'sen Coup de ros', there is no response.[95] As Le Clézio writes: 'Il s'en fout, des Coup de ros' (p. 309). What brings the two together, French and Creole, gentleman and tramp, is the beauty of the man's paintings, the choice of imagery transporting Dodo back to the landscape of Mauritius, to the mystical experiences of the early stages of his journey: it is clearly the imagination that lies at the core of Relation. The focus is on the transformative function of colour:

> il peint l'eau et le ciel mais ca ne ressemble pas à ce qu'il voit les couleurs sont fortes la mer très bleue et les voiles des bateaux rouges, les nuages blancs ou le ciel couleur de cyclone, parce qu ce qu'il peint c'est notre île, là-bas à l'autre bout de toutes les mers. (pp. 307–308)

The emphasis here is on the intensity of the colours. As the artist confirms: 'Tu vois, tu crois que j'exagère, mais ce sont les vraies couleurs, si tu fermes les yeux tu vois le violet partout, partout' (p. 308). Dodo confirms the veracity of this statement. It is indeed a fusional experience; there can be no distinction between art and life. As in the case of 'La fille aux cheveux bleus', colour is alive: 'Et moi je ne peux pas fermer les yeux mais je sais qu'il n'a pas menti. C'est violet partout (p. 308). Close affinities may be detected here with the texts of Yves Bonnefoy for whom intensity of colour in paintings likewise conveys a sense of immediacy, of pure presence, of 'ce monde d'avant les mots'.[96] Interestingly, for both writers, the painter can be seen to restore a relationship with nature that is being

95 The name Fe'sen refers to the side of Le Clézio's family that were originally plantation owners and whose fortune was built on slaves and coolies (p. 70).

96 See Yves Bonnefoy, 'la peinture du paysage', in Michel Collot, *La Pensée-paysage*, p. 212. The quotation is from Bonnefoy's *L'arbre au-delà des images*, published in 2003 by William Blake.

eroded in contemporary urban capitalist societies, in this case, those of Paris and Nice.

Indeed, a few lines later, the paintings and drawings are associated with pure movement and energy conveyed in images of the wind and the clouds, fuelling in Dodo the desire to return to his native land:

> J'aime bien regarder ses dessins et ses peintures dans son carnet, les paysages, les filaos penchés par le vent, les lagons, le ciel avec des petits nuages ronds parce qu'il n'y a qu'au pays Maurice que les nuages sont petits en troupeau de moutons, et ça me donne envie d'être là-bas, ça me donne envie de larmes mais j'ai les yeux secs et pour ça je dois les mouiller avec le bout de ma langue. (p. 309)

As we saw earlier in the evocation of Mauritius and of the Creole vision, the clouds are a primary image of Relation between earth and sky, life and death. Indeed, Dodo now repeats his desire to return, this time in images of flight and of the bird, images that have become central to his identity: 'Je veux m'envoler dans les images, je suis un oiseau qui s'échappe par la fenêtre' (p. 309).

As we learn later in the final chapter, 'Mon nom est Pe'sonne',[97] Dodo has been confined to a home for the destitute and the mentally disturbed, La Maison Blanche, where he can now live entirely in the world of the imagination: 'je peux imaginer Alma' (p. 327) and 'je suis dans mon île' (p. 328).[98] At the same time, Le Clézio draws attention to the power of colour as an agent of metamorphosis, the trigger for a fusional experience: it is the golden rays of the sun from the garden that transport Dodo back to the landscape of Mauritius: 'Je bois la couleur de l'île avec mes yeux' (p. 324).[99] As in the case of music, the boundaries between poetic expression and life are demolished: it can be described as an experience of material ecstasy,[100]

97 Dodo has become everyone, assuming the identities of the dead: there is a fusion between life and death, past and present.

98 In the garden of the asylum, Dodo can converse with the plants and the birds: 'Je regarde les plantes et les oiseaux, ils me parlent et moi je leur parle' (p. 326). We may be reminded here of Le Clézio's earlier essay *Haï*.

99 For a discussion of colour in painting, see also Emily McLaughlin, 'An Ecocritical Reading of Yves Bonnefoy's Punctual Colour Work', *French Studies*, Volume LXXI, Number 3, July 2017, pp. 359–361.

100 See Le Clézio's poetic essay *L'Extase matérielle*, p. 176.

a recovery of Dodo's Creole origins as embodied in the landscape of Alma and in the figure of his nurse Artémisia:

> C'est Alma, mon Alma, Alma des champs et des ruisseaux, des mares et des bois noirs, Alma dans mon cœur, Alma dans mon ventre. Tout le monde peut mourir, pikni, mais pas toi, Artémisia, pas toi. Je reste immobile dans le soleil d'or, les yeux levés vers l'intérieur de ma tête puisque je ne peux pas dormir, un jour mon âme va partir par un trou dans ma tête, pour aller au ciel où sont les étoiles. (p. 328)

Music

Le Clézio concludes his story with a return to the theme of music, as we have seen, a major structuring device of the book. We learn that it is New Year's Eve at the institution to which Dodo has been confined. Here the final paragraph opens with a focus on cosmic harmony, on the dance of life:

> Les jours et les nuits s'attachent, sans se casser, c'est un lent sac et ressac, un grand ballet qui emporte les gens, ceux d'Alma et ceux de l'asile, mon papa et Maman Laros, Yaya, Artémisia, et aussi Vicky, là-bas à l'autre bout de la mer. (p. 328)

This theme of the circle of life is strengthened in a reference to the fir tree that is planted in the hallway year after year. Furthermore, its needles are described as 'jaunes comme les dents de mon papa qui fume trop', suggesting communication between the living and the dead as illustrated much earlier in the figure of the mango tree (p. 65). Dodo is clearly giving expression to his Creole roots.

Indeed, the festivities accumulate in a celebration of song itself. In the concluding paragraphs of the story, Le Clézio highlights the theme of beauty, that is, the beauty of sound, a quality that has already emerged as a major catalyst for Relation and that lies at the very core of the living. Dodo is given permission to play and sing 'Auld Lang Syne' in Scottish, declaring, 'ce sont les mots les plus zolis, les plis doux de toutes les langues des hommes et des animaux' (p. 329). The focus here on the very sound of words may also call to mind the beauty of Ruben's poems mentioned at the very beginning of Dodo's visit to Paris.

Significantly, this performance is repeated every afternoon: the residents themselves burst into an ecstatic outpouring of sound. For Dodo, the song 'Auld Lang Syne' conjures up the image of his parents in the Saint-Jean cemetery, music here becoming a direct means of communication with the dead, with the invisible.[101] Time has indeed been regained. As Le Clézio writes:

> Tous les fous vaillants se réunissent dans le salon, je vais au piano, ils chantent avec moi les paroles dans la langue écossaise de ma grand-mère Beth et mon papa et même Maman Laros les entendent là où ils sont, ça doit leur faire chaud à l'âme. (p. 329)

Le Clézio concludes his book by presenting the reader with the words of the song in Scottish, conveying the life-affirming magical qualities of sound and of language:

> Ar oiche auld lang syne seo muid,
> Ar oiche auld lang syne
> Ag casadh amach anocht le brod
> *ar oiche auld lang syne.* (p. 329)

In spite of a very contemporary focus on the horrors of racism and of war, Le Clézio has chosen to end *The Story of Dodo* with the resurrection of life, with a celebration of the power of the imagination and of music. As Proust writes in reference to Vinteuil's septet, 'quelque chose comme un mystique chant du coq, un appel ineffable mais suraigu, de l'éternel matin'.[102]

101 The song may recall Proust's 'la petite phrase', described as 'un être surnaturel et pur qui passe en déroulant son message invisible'. See Marcel Proust, *À la Recherche du Temps Perdu* (Paris: Éditions Gallimard, 1954), p. 327. As Le Clézio has stated in an interview, 'La musique, ce n'est pas par hasard, est la forme d'art qui m'émeut le plus profondément jusqu'à me faire venir aux larmes. Peut-être parce qu'elle a faire avec l'invisible'. See 'Interview de Le Clézio: 'Il est temps de regarder vers le futur'/Association Des Lecteurs de J. M. G. Le Clézio (6 June 2020), <https://www.associationleclezio.com/interview-de-le-clezio-il-est-temps-de-regarder-vers-le-futur>.

102 For the second quotation from Proust, see Malcolm Bowie, *Proust among the Stars* (New York: HarperCollins, 1998), p. 79.

Conclusion

As noted earlier, Le Clézio's ecological vision is rooted in his own experiences of travel, of journeys through different cultures and different forms of life. A seminal role was played by the period he spent in the 1970s living amongst the Indigenous peoples of the Panamanian rainforest, 'une expérience qui a changé toute mes idées sur le monde et sur l'art'.[1] This experience is conveyed in his poetic essay *Haï* positing the existence of a universal intelligence or language, the expression of *le vivant*, of material energy. In a sharp critique of representational language, the essay celebrates the power of the imagination, highlighting the role of sensorial intensity in a communication with the non-human world, drawing attention at the same time to the act of metamorphosis, a rendering of Deleuze's concept of 'becoming' of 'becoming-animal'. Le Clézio has declared: 'J'écris pour essayer de savoir qui je suis.'[2] As we have seen, all four texts explore fusional relationships with the natural world as well as with the cosmic, a relationship that Le Clézio has described as mystical.

Indeed, Le Clézio's writing is distinguished by its highly poetic, almost magical nature. The border between language and life, text and world is shattered. A central role is played here by sound, by the very sound of words. Furthermore, it is music that lies at the very core of *le vivant*, bringing to the fore the interconnectedness of all forms of life. Attention is drawn to the act of repetition, to its crucial function in maintaining the cycle of life, all the more pertinent in the light of the ecological destruction we are now witnessing. This theme is aptly illustrated in the story of the spider, in the very act of spinning.

Importantly, Le Clézio's ecological vision is never divorced from everyday life or the political. A central theme of two of the books, *Bitna*,

1 See *La Fête chantée et autres essais de thème amérindien*, p. 9.

2 This statement was made in the interview with France Culture in 2008.

sous le ciel de Séoul and *Alma*, is that of interculturality or Relation, explored in the image of the bird, of 'becoming-bird'. In *Bitna,* set against the background of the Korean War (1950–1953), Salomé, under the hypnotic power of the narrator, undergoes a metamorphosis, 'glissait dans l'air avec le couple de pigeons' (p. 36), and feels herself flying from South Korea towards the North with messages of love and peace to members of the family who have been separated by war. In *The Story of Dodo*, the bird-like narrator is presented as in continual flight, developing a Relation with people from a variety of cultures, highlighting, as in *Bitna, sous le ciel de Séoul,* the political role of the imagination. Interestingly, Le Clézio has declared in an interview: 'Earth belongs to everyone, the right of passage has to exist. Birds fly over borders, why can't human beings?'

At the same time, Le Clézio draws attention to the fragility of this poetic vision, to the need for continual creative expression, for a series of 'becomings' in the face of ecological destruction, suffering and war. As he stated in his Nobel Prize speech in 2008: 'Agir, c'est ce que l'écrivain voudrait par-dessus tout. Agir plutôt que témoigner.'[3] In *Alma,* the contemporary Western city is associated with the pollution of the environment, in particular with that of water, the very source of life. In addition, there is a critique of the machine, a word that can also denote people, implying a total eradication of the human. Earlier, in *Bitna, sous le ciel de Séoul,* the narrator's detailed description of life in the city emerges as a comprehensive attack on a capitalist, materialist and egocentric society,[4] one with little concern for human suffering. Indeed, referring to the mortally ill Salomé, Le Clézio writes: 'Le monde qui est cruel et égoïste parce qu'il ne s'occupe pas d'elle. Le monde auquel elle ne manquera pas quand elle mourra' (p. 201).

Finally, within a focus on war in both *Bitna* and *Alma,* Le Clézio is bringing to the fore the dire consequences of humanism, of a concept of the self that is essentially abstract and that accords little value to the senses or to the imagination. Language, instead of being an act of creation, or the expression of the new, is purely referential, that is, the embodiment of power,

3 'Dans la forêt des paradoxes', September 2008. See Chapter 2, p. 14 for an earlier reference to the Nobel Prize speech.

4 See *Bitna, sous le ciel de Séoul,* Note 60, p. 17, where Le Clézio's gives his own damning critique of the city of Seoul.

of 'what has already been said and done', in the words of Wittgenstein. As we have seen, the narrative of *Bitna* is triggered by the war between North and South Korea, by the separation of families and communities. In *Alma*, a particularly striking example is afforded by the second half of *The Story of Dodo* where we are confronted with images of homeless child refugees roaming the streets of Paris, the victims of Western neocolonialism and oppression. Later, the racist murder of an Algerian war victim who has accompanied Dodo on his travels gives rise to one of the most powerful speeches Le Clézio has ever written, highlighting the enormous gap between rich and poor and the lack of any social concern.[5] Nevertheless, Dodo, constantly in movement, in transformation, refuses to become a prisoner of history: confined to an asylum, he sits in the garden where 'Je regarde les plantes et les oiseaux, ils me parlent et moi je leur parle' (p. 326). The story ends with a celebration of life and an assertion of the transformative power of the imagination and of music.

5 In an interview with Claude Cavallero, Le Clézio suggests that the novel(s) should help us understand our own history. See 'Les Marges et l'Origine: entretien avec J.M.G. Le Clézio' (*Europe* no. 954–958/janvier–février 2009), p. 31.

Bibliography

Ackerman, Jennifer, *The Genius of Birds* (London: Corsair, 2017).

Agamben, Giorgio, *Le Feu et le récit*, trans. Martin Rueff (Paris: Rivages, 2015)

——, *L'Ouvert: de l'homme et de l'animal* (London: Éditions Payot et Rivages, 2006).

Ameisen, Jean-Claude, *Sur Les Épaules de Darwin: Retrouver L'Aube* (Paris: Éditions Les Liens qui Liberent) (France Inter, 2014).

Arendt, Hannah, *Eichmann in Jerusalem: A Report on the Banality of Evil* (London: Penguin Classics, 2006).

Bachelard, Gaston, *L'Air et les songes, Essai sur l'imagination du mouvement 1943* (Paris: José Corti, 1944).

Badiou, Alain, *L'éthique* (Caen: NOUS, 2003).

——, *Wittgenstein's Antiphilosophy* (London: Verso, 2019).

Bailly, Jean-Christophe, *Le parti pris des animaux* (Paris: Christian Bourgeois, 2013).

——, *Le versant animal* (Paris: Bayard, 2007).

Balint, Adina, *Le processus de création dans l'oeuvre de J.M.G. Le Clézio* (Leiden: Brill, 2016).

Bennett, Jane, *New Materialisms Vibrant Matter: A Political Ecology of Things* (Durham, NC: Duke University Press, 2010).

Bergson, Henri, *La conscience et la vie* (1911), dir. Frédéric Worms (Paris: Presses Universitaires de France, 2011).

——, *L'Évolution créatrice, 1907* (Paris: Presses Universitaires de France 2011).

Blanchard, Émile, *Les Araignées* (Paris: Éditions VillaRrose, 2013).

Blanchot, Maurice, *L'Espace littéraire* (Paris: Éditions Gallimard, 1955).

Bowie, Malcolm, *Proust among the Stars* (New York: HarperCollins, 1998), p. 79.

Calvino, Italo, *Mr. Palomar* (London: Pan Books, 1986).

Cavallero, Claude, *Le Clézio, témoin du monde* (Paris: Éditions Calliopées, 2009).

——, '*Les Marges et L'Origine*: Entretien avec J.-M.G. Le Clézio', *Europe* no. 954–958/janvier–février 2009, pp. 29–38.

——, 'La tentation poétique de J.-M.G. Le Clézio', *Les Cahiers J.-M.G. Le Clézio*, no. 5/2012, pp. 9–12.

Césaire, Aimé, *Cahier d'un retour au pays natal* (Tarset: Bloodaxe Books, 1995).

——, *Discours sur le colonialisme* (Paris: Présence Africaine, 1955 and 2004).

Chamoiseau, Patrick, *Les Neuf Consciences du Malfini* (Paris: Éditions Gallimard, 2009).

Clark, Timothy, *The Value of Ecocriticism* (Cambridge: Cambridge University Press, 2019).

Collot, Michel, *La Pensée-paysage* (Actes Sud/ENS, 2011).

Connor, Steven, 'Making Flies Mean Something', in Mary Bryden, ed., *Beckett and Animals* (Cambridge: Cambridge University Press, 2015), pp. 139–152.

Deleuze, Gilles, *Critique et Clinique* (Paris: Éditions de Minuit, 1993).

——, *Différence et répétition* (Paris: Presses Universitaires de France, 1968).

——, *Proust et les signes* (Paris: Presses Universitaires De France, 1964).

Deleuze, Gilles, and Félix Guattari, *Mille Plateaux* (Paris: Éditions de Minuit, 1980).

——, *Qu'est-ce que la philosophie?* (Paris: Éditions de Minuit, 1991).

Deligny, Fernand, *The Arachnean and Other Texts*, trans. Drew S. Burk and Catherine Porter (Minneapolis: University of Minnesota Press, 2013).

Descola, Philippe, *L'Écologie des autres: l'anthropologie et la question de la nature* (Versailles: Quae, 2011).

——, *Par-delà nature et culture* (Paris: Gallimard, 2005).

Drouin, Jean-Marc, *Philosophie de l'insecte* (Paris: Éditions du Seuil, 2014).

Dubois, Philippe J., and Élise Rousseau, *Petite philosophie des oiseaux* (London: John Murray, 2021).

Dutton, Jacqueline, *Le Chercheur d'or et d'ailleurs: L'Utopie de J.M.G. Le Clézio* (Paris: L'Harmattan, 2003).

——, 'Entretien privé avec J.M.G. Le Clézio', le 29 mai 1997, in *Le Chercheur d'or et d'ailleurs: L'Utopie de J.M.G. Le Clézio* (Paris: L'Harmattan, 2003), 280–290.

Fontanille, Jacques, *Formes de vie* (Liège: Presses Universitaires de Liège, 2015).

——, 'Modes du sensible et syntaxe figurative', *Nouveaux actes semiotiques, 61–63* (1999), 1–68 (pp. 30–32).

de Fontenay, Elizabeth, *Quand un animal te regarde* (Paris: Éditions Gallimard, 2006).

Foucault, Michel, *Les Mots et les choses: une archéologie des sciences humaines* (Paris: Éditions Gallimard, 1966).

Gambert, Justyna, 'Entretien avec J.M.G. Le Clézio', *Les Cahiers J.-M.G. Le Clézio* no. 8/2015. 'Le Clézio et la Philosophie', pp. 65–74. This interview covers many of the core themes of Le Clézio's writing.

Ghosh, A., *The Nutmeg's Curse: Parables for a Planet in Crisis* (London: John Murray, 2021).

Glissant, Édouard, *La Cohée du Lamartin* (Paris: Éditions Gallimard, 2005).

——, *L'Intention Poétique* (Paris: Éditions du Seuil, 1969).

——, *La Poétique de le Relation* (Paris: Éditions Gallimard, 1990).

——, *Traité Du Tout Monde* (Paris: Éditions Gallimard, 1997).

Greimas, Algiradas-Julien, *De l'imperfectio*n (Perigueux: Fanlac, 1987).

Haraway, Donna J., *Staying with the Trouble* (Durham and London: Duke University Press, 2016).

——, *When Species Meet* (Minneapolis: University of Minnesota Press, 2008).
Huggan, Graham, and Helen Tiffin, *Postcolonial Ecocriticism: Literature, Animals and Environment* (New York: Routledge, 2015).
Jacobs, Dominique, *Splendeurs de l'araignée* (Paris: Klincksieck, 2018).
Kim, Hyeli, 'Les Paysages coréens dans *Bitna, sous le ciel de Séoul* comme une source de l'inspiration poétique', *Les Cahiers J.-M.G. Le Clézio* no. 13/2020: Signes et Songes d'Asie', pp. 78–86.
Klein, Naomi, *The Shock Doctrine: The Rise of Disaster Capitalism* (London: Penguin Books, 2007).
Kundera, Milan, *The Unbearable Lightness of Being*, trans. Michael Henry Heim (London: Faber and Faber, 1985).
Laplantine, François, *Penser le sensible* (Paris: Pocket Agora, 2018).
Latour, Bruno, *Face a Gaia* (Paris: La Découverte, 2015).
——, *Nous n'avons jamais été modernes: essai d'anthropologie symétrique* (Paris: La Découverte, 1997).
——, *Politiques de la nature: Comment faire entrer les sciences en démocracie* (Paris: La Découverte, 2015).
Le Clézio, J. M. G., *Ailleurs: Entretiens avec Jean-Louis Ezine* (Paris: Arléa, 1995).
——, *Alma* (Paris: Éditions Gallimard, 2017).
——, *Bita, sous le ciel de Séoul* (Paris: Éditions Stock, 2018).
——, *Le Chercheur d'or* (Paris: Éditions Gallimard, 1985).
——, *Cœur brûle et autres romances* (Paris: Gallimard, 2000).
——, Conférence Nobel (2008).
——, *Dans la forêt des paradoxes* (2008). Conference Nobel le 7 décembre 2008
——, *Désert* (Paris: Éditions Gallimard, 1980).
——, *L'extase matérielle* (Paris: Éditions Gallimard, Folio 1967).
——, *La fête chantée* (Paris: Éditions Gallimard, Le Promeneur, 1997).
——, *Les Géants* (Paris: Éditions Gallimard, 1973).
——, *Gens des nuages*, with Jémia Le Clézio (Paris: Stock, 1997).
——, *Haï* (Geneva: Skira, 1971).
——, *Histoire du pied et autres fantaisies* (Paris: Éditions Gallimard, 2011).
——, *L'inconnu sur la ter*re (Paris: Éditions Gallimard, 1978).
——, *Mondo et autres histoires* (Paris: Éditions Gallimard, 1978).
——, 'Nos vies d'araignées', in *Histoire du pied et d'autres fantaisies* (Paris: Éditions Gallimard, 2011), 11–346.
——, *Ourania* (Paris: Éditions Gallimard, 2006).
——, *Poisson d'Or* (Paris: Éditions Gallimard, 1996)
——, *La Procès-verb*al (Paris: Éditions Gallimard, Folio, 1963).
——, *La quarantaine* (Paris: Éditions Gallimard, 1995).
——, *Le Rêve Mexicain* (Paris: Éditions Gallimard, 1988).

——, *Révolutions* (Paris: Éditions Gallimard, 2003).

——, *Sirandanes*, with Jémia Le Clézio (Paris: Éditions Seghers, 1990).

Léger, Thierry, 'L'Asie et l'interculturel dans l'oeuvre de Le Clézio', *Les Cahiers J.-M.G .Le Clézio* no. 13/2020: 'Signes et Songes d'Asie', pp. 119–132.

Leopold, Aldo, *A Sand County Almanach* (New York: The Library of America, 2013).

Lévi-Strauss, Claude, *Structural Anthropology 2* (London: Penguin Books, 1994).

Maciel, Maria Esther, 'Interview with Dominique Lestel', *Contemporary French and Francophone Studies* 16/5, 711–719.

Maffie, James, *Aztec Philosophy: Understanding a World in Motion* (Boulder: University Press of Colorado, 2013).

Merleau-Ponty, Maurice, *Nature: Course Notes from Le Collège de France*, ed. Dominique Séglard, trans. Robert Vallier (Evanston, IL: Northwestern University Press, 2003).

——, *La Nature: notes, cours du Collège de France*, ed. Dominique Ségland (Paris: Seuil, 1994).

——, *Le Visible et l'invisible* (Paris: Éditions Gallimard, 1974).

Nancy, Jean-Luc, *The Birth to Presence*, trans. Brian Holmes and others (Stanford: Stanford University Press, 1993).

——, *Listening*, trans. Charlotte Mandell (New York: Fordham University Press, 2006).

Nielipowicz, *Natalia*, *Quelques Traces De L'Écophénoménologie dans Le Chercheur d'Or'*, *Les Cahiers J.-M.G. Le Clézio* no. 10/2017: 'Habiter la Terre', pp. 57–67.

Parret, Hermann, *Épiphanies de la présence* (Paris: Presses Universitaires de Limoges, 2006).

Peirce, Charles S., *Collected Papers of Charles Sanders Peirce*, eds. Charles Hartshorne and Paul Weiss (Cambridge, MA: Harvard University Press, 1931–1958, 1V 1933).

——, 'Prolegomena to an Apology for Pragmatism', in *Collected Papers of Charles Sanders Peirce*, eds. Charles Hartshorne and Paul Weiss (Cambridge, MA: Harvard University Press, 1931–1958, Vol. I (1933), pp. 530–572.

Pierron, Jean-Philippe, 'L'écrivain et l'écologiste: Le Clézio lecteur d'Aldo Leopold', in *Les Puissances de l'Imagination* (Paris: Les Éditions du Cerf, 2012), 89–95.

Pouydebot, Emmanuelle, *L'Intelligence animale* (Paris: Odile Jacob, 2017).

Premat, Christophe, 'Retrouver les voix autochtones: *l'hommage de Jean-Marie Gustave Le Clézio à Rita Mestokosho*', *Les Cahiers J.-M.G. Le Clézio* no. 7/2014: 'Le gout des langues, les langues à l'oeuvre', pp. 119–33.

——, 'Les Vibrations de la Conscience Minéralisée' dans *L'inconnu sur la terre* de J. M. G. Le Clézio', *Les Cahiers J.-M.G. Le Clézio* no. 12/2019: 'Corps', pp. 45–55.

Proust, Marcel, *À la Recherche du Temps Perdu* (Paris: Éditions Gallimard, 1954).

Rilke, Rainer Maria, *La Huitième Élegie de Duino* in Maurice Blanchot, *L'espace littéraire* (Paris: Éditions Gallimard, 1995).
Rousseau, Jean-Jacques, *Discours sur l'origine de l'inégalité parmi les hommes*, ed. Jean Starobinski (Paris: Éditions Gallimard, 1995).
Roussel-Gillet, Isabelle, 'En dialogue avec Jean-Marie Gustave Le Clézio', *Roman 20–50: Revue d'étude du roman du XXe siècle*, no. 55/2013, pp. 131–146. This includes a discussion of the crucial ecological role of insects.
Rueckert, William, 'Literature and Ecology', in Cheryll Glotfelty and Harold From, eds, *The Ecocriticism Reader* (Athens: University of Georgia Press, 1996) 108–112.
Schopenhauer, Arthur, *Parega et Paralipomena: Short Philosophical Essays* (The 1851 Cambridge Edition of the Works of Schopenhauer, 2017) http://www.septentrion.com/fr/livre/?GCOI=27574100527550.
——, *The World as Will and Representation*, trans. E. F. J. Payne (New York: Dover, 1969).
Simon, Anne, 'Zoopoétique', *Revue des Sciences Humaines* 328 4/201.
Suberchicot, Alain, *Littérature Et Environnement: Pour une écocritique comparée* (Paris: Honoré Champion, 2012).
Uexküll, Jacob von. *Milieu animal et milieu humain 1956* (Paris: Éditions Payot et Rivagges, 2010).
Van Der Drift, Martha, 'Entretien avec J.M.G. Le Clézio', *Les Cahiers J.-M.G. Le Clézio* no. 5/2012: 'La Tentation poétique', pp. 162–164. Here, the theme of interculturality is explored with reference to Issa Asgarally, Édouard Glissant and *créolité* (pp. 162–165). Later, the collection of stories titled *Histoire du pied et autres fantaisies* becomes an object of discussion (p. 168).
Wald, Aliocha, 'Entretien avec Philippe Descola', in *Panorama de la pensée d'aujourd'hui* (Paris: Fayard, 2016), 712–34.
Wiseman, Boris, *Lévi-Strauss, Anthropology and Aesthetics* (Cambridge: Cambridge University Press, 2007).
Wittgenstein, Ludwig, *Tractatus Logico-Philosophicus* (London and New York: Routledge & Kegan Paul, 1961).
Zwang, Lu, 'Entretien de Lu Zwang avec J.M.G Le Clézio sur Les Philosophies Orientales et Leurs Rapports avec L'Occident', *Les Cahiers J.-M.G. Le Clézio* no. 8/2015: 'Le Clézio Et La Philosophie', pp. 87–101. Le Clézio states that European philosophers have done nothing to prevent war or racism (p. 98).

Index

Modern French Identities

Edited by Jean Khalfa

This series aims to publish monographs, editions or collections of papers based on recent research into modern French Literature. It welcomes contributions from academics, researchers and writers worldwide and in British and Irish universities in particular.

Modern French Identities focuses on the French and Francophone writing of the twentieth and twenty-first centuries, whose formal experiments and revisions of genre have combined to create an entirely new set of literary forms, from the thematic autobiographies of Michel Leiris and Bernard Noël to the magic realism of French Caribbean writers.

The idea that identities are constructed rather than found, and that the self is an area to explore rather than a given pretext, runs through much of modern French literature, from Proust, Gide, Apollinaire and Césaire to Barthes, Duras, Kristeva, Glissant, Germain and Roubaud.

This series explores the turmoil in ideas and values expressed in the works of theorists like Lacan, Irigaray, Foucault, Fanon, Deleuze and Bourdieu and traces the impact of current theoretical approaches – such as gender and sexuality studies, de/coloniality, intersectionality, and ecocriticism – on the literary and cultural interpretation of the self.

The series publishes studies of individual authors and artists, comparative studies, and interdisciplinary projects and welcomes research on autobiography, cinema, fiction, poetry and performance art and/or the intersections between them.

Volume 1	Victoria Best & Peter Collier (eds): Powerful Bodies. Performance in French Cultural Studies. 220 pages. 1999. ISBN 3-906762-56-4 / US-ISBN 0-8204-4239-9
Volume 2	Julia Waters: Intersexual Rivalry. A 'Reading in Pairs' of Marguerite Duras and Alain Robbe-Grillet. 228 pages. 2000. ISBN 3-906763-74-9 / US-ISBN 0-8204-4626-2
Volume 3	Sarah Cooper: Relating to Queer Theory. Rereading Sexual Self-Definition with Irigaray, Kristeva, Wittig and Cixous. 231 pages. 2000. ISBN 3-906764-46-X / US-ISBN 0-8204-4636-X
Volume 4	Julia Prest & Hannah Thompson (eds): Corporeal Practices. (Re) figuring the Body in French Studies. 166 pages. 2000. ISBN 3-906764-53-2 / US-ISBN 0-8204-4639-4
Volume 5	Victoria Best: Critical Subjectivities. Identity and Narrative in the Work of Colette and Marguerite Duras. 243 pages. 2000. ISBN 3-906763-89-7 / US-ISBN 0-8204-4631-9
Volume 6	David Houston Jones: The Body Abject: Self and Text in Jean Genet and Samuel Beckett. 213 pages. 2000. ISBN 3-906765-07-5 / US-ISBN 0-8204-5058-8
Volume 7	Robin MacKenzie: The Unconscious in Proust's A la recherche du temps perdu. 270 pages. 2000. ISBN 3-906758-38-9 / US-ISBN 0-8204-5070-7
Volume 8	Rosemary Chapman: Siting the Quebec Novel. The Representation of Space in Francophone Writing in Quebec. 282 pages. 2000. ISBN 3-906758-85-0 / US-ISBN 0-8204-5090-1
Volume 9	Gill Rye: Reading for Change. Interactions between Text Identity in Contemporary French Women's Writing (Baroche, Cixous, Constant). 223 pages. 2001. ISBN 3-906765-97-0 / US-ISBN 0-8204-5315-3
Volume 10	Jonathan Paul Murphy: Proust's Art. Painting, Sculpture and Writing in A la recherche du temps perdu. 248 pages. 2001. ISBN 3-906766-17-9 / US-ISBN 0-8204-5319-6
Volume 11	Julia Dobson: Hélène Cixous and the Theatre. The Scene of Writing. 166 pages. 2002. ISBN 3-906766-20-9 / US-ISBN 0-8204-5322-6

Volume 12 Emily Butterworth & Kathryn Robson (eds): Shifting Borders. Theory and Identity in French Literature.
226 pages. 2001. ISBN 3-906766-86-1 / US-ISBN 0-8204-5602-0

Volume 13 Victoria Korzeniowska: The Heroine as Social Redeemer in the Plays of Jean Giraudoux.
144 pages. 2001. ISBN 3-906766-92-6 / US-ISBN 0-8204-5608-X

Volume 14 Kay Chadwick: Alphonse de Châteaubriant: Catholic Collaborator.
327 pages. 2002. ISBN 3-906766-94-2 / US-ISBN 0-8204-5610-1

Volume 15 Nina Bastin: Queneau's Fictional Worlds.
291 pages. 2002. ISBN 3-906768-32-5 / US-ISBN 0-8204-5620-9

Volume 16 Sarah Fishwick: The Body in the Work of Simone de Beauvoir.
284 pages. 2002. ISBN 3-906768-33-3 / US-ISBN 0-8204-5621-7

Volume 17 Simon Kemp & Libby Saxton (eds): Seeing Things. Vision, Perception and Interpretation in French Studies.
287 pages. 2002. ISBN 3-906768-46-5 / US-ISBN 0-8204-5858-9

Volume 18 Kamal Salhi (ed.): French in and out of France. Language Policies, Intercultural Antagonisms and Dialogue. 487 pages. 2002. ISBN 3-906768-47-3 / US-ISBN 0-8204-5859-7

Volume 19 Genevieve Shepherd: Simone de Beauvoir's Fiction. A Psychoanalytic Rereading.
262 pages. 2003. ISBN 3-906768-55-4 / US-ISBN 0-8204-5867-8

Volume 20 Lucille Cairns (ed.): Gay and Lesbian Cultures in France.
290 pages. 2002. ISBN 3-906769-66-6 / US-ISBN 0-8204-5903-8

Volume 21 Wendy Goolcharan-Kumeta: My Mother, My Country. Reconstructing the Female Self in Guadeloupean Women's Writing.
236 pages. 2003. ISBN 3-906769-76-3 / US-ISBN 0-8204-5913-5

Volume 22 Patricia O'Flaherty: Henry de Montherlant (1895–1972). A Philosophy of Failure.
256 pages. 2003. ISBN 3-03910-013-0 / US-ISBN 0-8204-6282-9

Volume 23 Katherine Ashley (ed.): Prix Goncourt, 1903–2003: essais critiques.
205 pages. 2004. ISBN 3-03910-018-1 / US-ISBN 0-8204-6287-X

Volume 24 Julia Horn & Lynsey Russell-Watts (eds): Possessions. Essays in French Literature, Cinema and Theory.
223 pages. 2003. ISBN 3-03910-005-X / US-ISBN 0-8204-5924-0

Volume 25 Steve Wharton: Screening Reality.
French Documentary Film during the German Occupation.
252 pages. 2006. ISBN 3-03910-066-1 / US-ISBN 0-8204-6882-7

Volume 26 Frédéric Royall (ed.): Contemporary French Cultures and Societies.
421 pages. 2004. ISBN 3-03910-074-2 / US-ISBN 0-8204-6890-8

Volume 27 Tom Genrich: Authentic Fictions.
Cosmopolitan Writing of the Troisième République, 1908–1940.
288 pages. 2004. ISBN 3-03910-285-0 / US-ISBN 0-8204-7212-3

Volume 28 Maeve Conrick & Vera Regan: French in Canada.
Language Issues.
186 pages. 2007. ISBN 978-3-03-910142-9

Volume 29 Kathryn Banks & Joseph Harris (eds): Exposure.
Revealing Bodies, Unveiling Representations.
194 pages. 2004. ISBN 3-03910-163-3 / US-ISBN 0-8204-6973-4

Volume 30 Emma Gilby & Katja Haustein (eds): Space.
New Dimensions in French Studies.
169 pages. 2005. ISBN 3-03910-178-1 / US-ISBN 0-8204-6988-2

Volume 31 Rachel Killick (ed.): Uncertain Relations.
Some Configurations of the 'Third Space' in Francophone Writings of the Americas and of Europe.
258 pages. 2005. ISBN 3-03910-189-7 / US-ISBN 0-8204-6999-8

Volume 32 Sarah F. Donachie & Kim Harrison (eds): Love and Sexuality.
New Approaches in French Studies.
194 pages. 2005. ISBN 3-03910-249-4 / US-ISBN 0-8204-7178-X

Volume 33 Michaël Abecassis: The Representation of Parisian Speech in the Cinema of the 1930s.
409 pages. 2005. ISBN 3-03910-260-5 / US-ISBN 0-8204-7189-5

Volume 34 Benedict O'Donohoe: Sartre's Theatre: Acts for Life.
301 pages. 2005. ISBN 3-03910-250-X / US-ISBN 0-8204-7207-7

Volume 35 Moya Longstaffe: The Fiction of Albert Camus. A Complex Simplicity.
300 pages. 2007. ISBN 3-03910-304-0 / US-ISBN 0-8204-7229-8

Volume 36 Arnaud Beaujeu: Matière et lumière dans le théâtre de Samuel Beckett: Autour des notions de trivialité, de spiritualité et d' « autre-là ».
377 pages. 2010. ISBN 978-3-0343-0206-8

Volume 37 Shirley Ann Jordan: Contemporary French Women's Writing: Women's Visions, Women's Voices, Women's Lives.
308 pages. 2005. ISBN 3-03910-315-6 / US-ISBN 0-8204-7240-9

Volume 38 Neil Foxlee: Albert Camus's 'The New Mediterranean Culture': A Text and its Contexts.
349 pages. 2010. ISBN 978-3-0343-0207-4

Volume 39 Michael O'Dwyer & Michèle Raclot: Le Journal de Julien Green: Miroir d'une âme, miroir d'un siècle.
289 pages. 2005. ISBN 3-03910-319-9

Volume 40 Thomas Baldwin: The Material Object in the Work of Marcel Proust.
188 pages. 2005. ISBN 3-03910-323-7 / US-ISBN 0-8204-7247-6

Volume 41 Charles Forsdick & Andrew Stafford (eds): The Modern Essay in French: Genre, Sociology, Performance.
296 pages. 2005. ISBN 3-03910-514-0 / US-ISBN 0-8204-7520-3

Volume 42 Peter Dunwoodie: Francophone Writing in Transition. Algeria 1900–1945.
339 pages. 2005. ISBN 3-03910-294-X / US-ISBN 0-8204-7220-4

Volume 43 Emma Webb (ed.): Marie Cardinal: New Perspectives.
260 pages. 2006. ISBN 3-03910-544-2 / US-ISBN 0-8204-7547-5

Volume 44 Jérôme Game (ed.): Porous Boundaries: Texts and Images in Twentieth-Century French Culture.
164 pages. 2007. ISBN 978-3-03910-568-7

Volume 45 David Gascoigne: The Games of Fiction: Georges Perec and Modern French Ludic Narrative.
327 pages. 2006. ISBN 3-03910-697-X / US-ISBN 0-8204-7962-4

Volume 46 Derek O'Regan: Postcolonial Echoes and Evocations: The Intertextual Appeal of Maryse Condé.
329 pages. 2006. ISBN 3-03910-578-7

Volume 47 Jennifer Hatte: La langue secrète de Jean Cocteau: la mythologie personnelle du poète et l'histoire cachée des Enfants terribles.
332 pages. 2007. ISBN 978-3-03910-707-0

Volume 48 Loraine Day: Writing Shame and Desire: The Work of Annie Ernaux.
315 pages. 2007. ISBN 978-3-03910-275-4

Volume 61 Lucy Bolton, Gerri Kimber, Ann Lewis and Michael Seabrook (eds): Framed!: Essays in French Studies.
235 pages. 2007. ISBN 978-3-03911-043-8

Volume 62 Lorna Milne and Mary Orr (eds): Narratives of French Modernity: Themes, Forms and Metamorphoses. Essays in Honour of David Gascoigne.
365 pages. 2011. ISBN 978-3-03911-051-3

Volume 63 Ann Kennedy Smith: Painted Poetry: Colour in Baudelaire's Art Criticism.
253 pages. 2011. ISBN 978-3-03911-094-0

Volume 64 Sam Coombes: The Early Sartre and Marxism.
330 pages. 2008. ISBN 978-3-03911-115-2

Volume 65 Claire Lozier: De l'abject et du sublime: Georges Bataille, Jean Genet, Samuel Beckett.
327 pages. 2012. ISBN 978-3-0343-0724-6

Volume 66 Charles Forsdick and Andy Stafford (eds): La Revue: The Twentieth- Century Periodical in French.
379 pages. 2013. ISBN 978-3-03910-947-0

Volume 67 Alison S. Fell (ed.): French and francophone women facing war / Les femmes face à la guerre.
301 pages. 2009. ISBN 978-3-03911-332-3

Volume 68 Elizabeth Lindley and Laura McMahon (eds): Rhythms: Essays in French Literature, Thought and Culture. 238 pages. 2008. ISBN 978-3-03911-349-1

Volume 69 Georgina Evans and Adam Kay (eds): Threat: Essays in French Literature, Thought and Visual Culture.
248 pages. 2010. ISBN 978-3-03911-357-6

Volume 70 John McCann: Michel Houellebecq: Author of our Times.
229 pages. 2010. ISBN 978-3-03911-373-6

Volume 71 Jenny Murray: Remembering the (Post)Colonial Self: Memory and Identity in the Novels of Assia Djebar.
258 pages. 2008. ISBN 978-3-03911-367-5

Volume 72 Susan Bainbrigge: Culture and Identity in Belgian Francophone Writing: Dialogue, Diversity and Displacement.
230 pages. 2009. ISBN 978-3-03911-382-8

Volume 73 Maggie Allison and Angela Kershaw (eds): Parcours de femmes: Twenty Years of Women in French.
313 pages. 2011. ISBN 978-3-0343-0208-1

Volume 74 Jérôme Game: Poetic Becomings: Studies in Contemporary French Literature.
263 pages. 2011. ISBN 978-3-03911-401-6

Volume 75 Elodie Laügt: L'Orient du signe: Rêves et dérives chez Victor Segalen, Henri Michaux et Emile Cioran.
242 pages. 2008. ISBN 978-3-03911-402-3

Volume 76 Suzanne Dow: Madness in Twentieth-Century French Women's Writing: Leduc, Duras, Beauvoir, Cardinal, Hyvrard.
217 pages. 2009. ISBN 978-3-03911-540-2

Volume 77 Myriem El Maïzi: Marguerite Duras ou l'écriture du devenir.
228 pages. 2009. ISBN 978-3-03911-561-7

Volume 78 Claire Launchbury: Music, Poetry, Propaganda: Constructing French Cultural Soundscapes at the BBC during the Second World War.
223 pages. 2012. ISBN 978-3-0343-0239-5

Volume 79 Jenny Chamarette and Jennifer Higgins (eds): Guilt and Shame: Essays in French Literature, Thought and Visual Culture.
231 pages. 2010. ISBN 978-3-03911-563-1

Volume 80 Vera Regan and Caitríona Ní Chasaide (eds): Language Practices and Identity Construction by Multilingual Speakers of French L2: The Acquisition of Sociostylistic Variation.
189 pages. 2010. ISBN 978-3-03911-569-3

Volume 81 Margaret-Anne Hutton (ed.): Redefining the Real: The Fantastic in Contemporary French and Francophone Women's Writing.
294 pages. 2009. ISBN 978-3-03911-567-9

Volume 82 Elise Hugueny-Léger: Annie Ernaux, une poétique de la transgression.
269 pages. 2009. ISBN 978-3-03911-833-5

Volume 83 Peter Collier, Anna Magdalena Elsner and Olga Smith (eds): Anamnesia: Private and Public Memory in Modern French Culture. 359 pages. 2009. ISBN 978-3-03911-846-5

Volume 84 Adam Watt (ed./éd.): Le Temps retrouvé Eighty Years After/80 ans après: Critical Essays/Essais critiques.
349 pages. 2009. ISBN 978-3-03911-843-4

Volume 85 Louise Hardwick (ed.): New Approaches to Crime in French Literature, Culture and Film.
237 pages. 2009. ISBN 978-3-03911-850-2

Volume 86 Emmanuel Godin and Natalya Vince (eds): France and the Mediterranean: International Relations, Culture and Politics. 372 pages. 2012. ISBN 978-3-0343-0228-9

Volume 87 Amaleena Damlé and Aurélie L'Hostis (eds): The Beautiful and the Monstrous: Essays in French Literature, Thought and Culture.
237 pages. 2010. ISBN 978-3-03911-900-4

Volume 88 Alistair Rolls (ed.): Mostly French: French (in) Detective Fiction.
212 pages. 2009. ISBN 978-3-03911-957-8

Volume 89 Bérénice Bonhomme: Claude Simon: une écriture en cinéma.
359 pages. 2010. ISBN 978-3-03911-983-7

Volume 90 Barbara Lebrun and Jill Lovecy (eds): Une et divisible? Plural Identities in Modern France.
258 pages. 2010. ISBN 978-3-0343-0123-7

Volume 91 Pierre-Alexis Mével & Helen Tattam (eds): Language and its Contexts/ Le Langage et ses contextes: Transposition and Transformation of Meaning?/ Transposition et transformation du sens?
272 pages. 2010. ISBN 978-3-0343-0128-2

Volume 92 Alistair Rolls and Marie-Laure Vuaille-Barcan (eds): Masking Strategies: Unwrapping the French Paratext.
202 pages. 2011. ISBN 978-3-0343-0746-8

Volume 93 Michaël Abecassis et Gudrun Ledegen (éds): Les Voix des Français Volume 1: à travers l'histoire, l'école et la presse. 372 pages. 2010. ISBN 978-3-0343-0170-1

Volume 94 Michaël Abecassis et Gudrun Ledegen (éds): Les Voix des Français Volume 2: en parlant, en écrivant.
481 pages. 2010. ISBN 978-3-0343-0171-8

Volume 95 Manon Mathias, Maria O'Sullivan and Ruth Vorstman (eds): Display and Disguise.
237 pages. 2011. ISBN 978-3-0343-0177-0

Volume 96 Charlotte Baker: Enduring Negativity: Representations of Albinism in the Novels of Didier Destremau, Patrick Grainville and Williams Sassine.
226 pages. 2011. ISBN 978-3-0343-0179-4

Volume 97 Florian Grandena and Cristina Johnston (eds): New Queer Images: Representations of Homosexualities in Contemporary Francophone Visual Cultures.
246 pages. 2011. ISBN 978-3-0343-0182-4

Volume 98 Florian Grandena and Cristina Johnston (eds): Cinematic Queerness: Gay and Lesbian Hypervisibility in Contemporary Francophone Feature Films.
354 pages. 2011. ISBN 978-3-0343-0183-1

Volume 99 Neil Archer and Andreea Weisl-Shaw (eds): Adaptation: Studies in French and Francophone Culture.
234 pages. 2012. ISBN 978-3-0343-0222-7

Volume 100 Peter Collier et Ilda Tomas (éds): Béatrice Bonhomme: le mot, la mort, l'amour.
437 pages. 2013. ISBN 978-3-0343-0780-2

Volume 101 Helena Chadderton: Marie Darrieussecq's Textual Worlds: Self, Society, Language.
170 pages. 2012. ISBN 978-3-0343-0766-6

Volume 102 Manuel Bragança: La crise allemande du roman français, 1945–1949: la représentation des Allemands dans les best-sellers de l'immédiat après-guerre.
220 pages. 2012. ISBN 978-3-0343-0835-9

Volume 103 Bronwen Martin: The Fiction of J. M. G. Le Clézio: A Postcolonial Reading.
199 pages. 2012. ISBN 978-3-0343-0162-6

Volume 104 Hugues Azérad, Michael G. Kelly, Nina Parish et Emma Wagstaff (éds): Chantiers du poème: prémisses et pratiques de la création poétique moderne et contemporaine.
374 pages. 2013. ISBN 978-3-0343-0800-7

Volume 105 Franck Dalmas: Lectures phénoménologiques en littérature française: de Gustave Flaubert à Malika Mokeddem.
253 pages. 2012. ISBN 978-3-0343-0727-7

Volume 106 Béatrice Bonhomme, Aude Préta-de Beaufort et Jacques Moulin (éds): Dans le feuilletage de la terre: sur l'œuvre poétique de Marie- Claire Bancquart.
533 pages. 2013. ISBN 978-3-0343-0721-5

Volume 107 Claire Bisdorff et Marie-Christine Clemente (éds): Le Cœur dans tous ses états: essais sur la littérature et l'art français.
230 pages. 2013. ISBN 978-3-0343-0711-6

Volume 108 Michaël Abecassis et Gudrun Ledegen (éds): Écarts et apports des médias francophones: lexique et grammaire.
300 pages. 2013. ISBN 978-3-0343-0882-3

Volume 109 Allison and Imogen Long (eds): Women Matter / Femmes Matière: French and Francophone Women and the Material World. 273 pages. 2013. ISBN 978-3-0343-0788-8

Volume 110 Fabien Arribert-Narce et Alain Ausoni (éds): L'Autobiographie entre autres: écrire la vie aujourd'hui.
221 pages. 2013. ISBN 978-3-0343-0858-8

Volume 111 Leona Archer and Alex Stuart (eds): Visions of Apocalypse: Representations of the End in French Literature and Culture.
266 pages. 2013. ISBN 978-3-0343-0921-9

Volume 112 Simona Cutcan: Subversion ou conformisme ? La différence des sexes dans l'œuvre d'Agota Kristof.
264 pages. 2014. ISBN 978-3-0343-1713-9

Volume 113 Owen Heathcote: From Bad Boys to New Men ? Masculinity, Sexuality and Violence in the Work of Éric Jourdan.
279 pages. 2014. ISBN 978-3-0343-0736-9

Volume 114 Ilda Tomas: Arc-en-ciel: études sur divers poètes.
234 pages. 2014. ISBN 978-3-0343-0975-2

Volume 115 Lisa Jeschke and Adrian May (eds): Matters of Time: Material Temporalities in Twentieth-Century French Culture.
314 pages. 2014. ISBN 978-3-0343-1796-2

Volume 116 Crispin T. Lee: Haptic Experience in the Writings of Georges Bataille, Maurice Blanchot and Michel Serres.
316 pages. 2014. ISBN 978-3-0343-1791-7

Volume 117 Ashwiny O. Kistnareddy: Locating Hybridity: Creole, Identities and Body Politics in the Novels of Ananda Devi.
208 pages. 2015. ISBN 978-3-0343-1814-3

Volume 118 Michaël Abecassis et Gudrun Ledegen (éds): De la genèse de la langue à Internet: variations dans les formes, les modalités et les langues en contact.
278 pages. 2015. ISBN 978–3-0343-1798-6

Volume 119 Peter D. Tame: Isotopias: Places and Spaces in French War Fiction of the Twentieth and Twenty-First Centuries.
592 pages. 2015. ISBN 978–3-0343-0837-3

Volume 120 Daniel A. Finch-Race et Jeff Barda (éds): Textures: Processus et événements dans la création poétique moderne et contemporaine. 242 pages. 2015. ISBN 978–3-0343-1898-3

Volume 121 Hélène Sicard-Cowan: Vivre ensemble: éthique de l'imitation dans la littérature et le cinéma de l'immigration en France (1986– 2005). 149 pages. 2016. ISBN 978–3-0343-1944-7

Volume 122 Mercedes Montoro Araque et Carmen Alberdi Urquizu (éds): L'entre- deux imaginaire: corps et création interculturels.
216 pages. 2016. ISBN 978–3-0343-1926-3

Volume 123 Maureen A. Ramsden: Crossing Borders: The Interrelation of Fact and Fiction in Historical Works, Travel Tales, Autobiography and Reportage.
191 pages. 2016. ISBN 978–3-0343-1995-9

Volume 124 Jean Khalfa: Poetics of the Antilles: Poetry, History and Philosophy in the Writings of Perse, Césaire, Fanon and Glissant.
388 pages. 2017. ISBN 978–3-0343-0895-3

Volume 125 Mathilde Poizat-Amar: L'Eclat du voyage: Blaise Cendrars, Victor Segalen, Albert Londres.
252 pages. 2017. ISBN 978–1-78707-296-1

Volume 126 Philippe Willemart: L'Univers de la création littéraire: Dans la chambre noire de l'écriture: Hérodias de Flaubert.
160 pages. 2017. ISBN 978–1-78707-458-3

Volume 127 Margaret Atack, Alison S. Fell, Diana Holmes and Imogen Long (eds): French Feminisms 1975 and After: New Readings, New Texts. 276 pages. 2018. ISBN 978-3-0343-2209-6

Volume 128 Matt Phillips and Tomas Weber (eds): Parasites: Exploitation and Interference in French Thought and Culture.
284 pages. 2018. ISBN 978-3-0343-2266-9

Volume 129 Zoe Angelis and Blake Gutt (eds): Stains / Les taches: Communication and Contamination in French and Francophone Literature and Culture.
274 pages. 2019. ISBN 978-1-78707-443-9

Volume 130 Michaël Abecassis avec Marcelline Block, Gudrun Ledegen et Maribel Peñalver Vicea (éds): Le Grain de la voix dans le monde anglophone et francophone.
332 pages. 2019. ISBN 978-1-78874-107-1

Volume 131 Philippe Willemart: L'écriture à l'ère de l'indétermination: Études sur la critique génétique, la psychanalyse et la littérature.
232 pages. 2019. ISBN 978–1-78874-631-1

Volume 132 Augustin Voegele: De l'unanimisme au fantastique: Jules Romains devant l'extraordinaire.
382 pages. 2019. ISBN 978-1-78874-513-0

Volume 133 Maggie Allison, Elliot Evans and Carrie Tarr (eds): Plaisirs de femmes: Women, Pleasure and Transgression in French Literature and Culture.
278 pages. 2019. ISBN 978-1-78874-383-9

Volume 134 Aaron Prevots: Bernard Vargaftig: Gestures toward the Sacred.
144 pages. 2019. ISBN 978-1-78997-357-0

Volume 135 Susie Cronin, Sofia Ropek Hewson and Cillian Ó Fathaigh (eds): #NousSommes: Collectivity and the Digital in French Thought and Culture. 184 pages. 2020. ISBN 978-1-78874-767-7

Volume 136 Anaïs Stampfli: La coprésence de langues dans le roman antillais contemporain. 456 pages. 2020. ISBN 978-1-78874-578-9

Volume 137 Philippe Willemart: Les mécanismes de la création littéraire: Lecture, écriture, génétique et psychanalyse. 214 pages. 2020. ISBN 978-1-78997-737-0

Volume 138 Maureen A. Ramsden: The Evolution of Proust's 'Combray': A Genetic Study.
152 pages. 2020. ISBN 978-1-78997-785-1

Volume 139 Charlène Clonts: Gherasim Luca: Texte, image, son.
454 pages. 2020. ISBN 978-1-78997-916-9

Volume 140 Polly Galis, Maria Tomlinson and Antonia Wimbush (eds): Queer(y)ing Bodily Norms in Francophone Culture.
304 pages. 2021. ISBN 978-1-78997-514-7

Volume 141 Natalie Berkman: OuLiPo and the Mathematics of Literature.
340 pages. 2022. ISBN 978-1-78997-780-6

Volume 142 David Gascoigne and Ana de Medeiros (eds): Marie Nimier: Le Sujet et ses écritures / The Self in the Web of Language. 312 pages. 2021. ISBN 978-1-80079-195-4

Volume 143 Carole Bourne-Taylor and Sara-Louise Cooper (eds): Variations on the Ethics of Mourning in Modern Literature in French.
332 pages. 2022. ISBN 978-1-78997-273-3

Volume 144 Sophie Large et Flora Valadié (éds): Le Fanon des artistes. Perspectives transaméricaines.
288 pages. 2022. ISBN 978-1-80079-630-0

Volume 145 Ashwiny O. Kistnareddy and Alice Roullière (eds): Catching up with Time: Belatedness and Anachronies in Francophone Literature and Culture.
258 pages. 2022. ISBN 978-1-80079-337-8

Volume 146 Alexandra Cătană: Le corps de la lettre: La quête de la totalité dans la poétique d'Isidore Isou
404 pages. 2023. ISBN 978-1-80079-706-2

Volume 147 Bronwen Martin: The Ecological Vision of J.M.G. Le Clézio
170 pages. 2024. ISBN 978-1-80374-067-6

Volume 148 Thea Petrou: Form and Love in the Poetry of Jacques Roubaud
Forthcoming. 2024. ISBN 978-1-80079-265-4

www.ingramcontent.com/pod-product-compliance
Ingram Content Group UK Ltd.
Pitfield, Milton Keynes, MK11 3LW, UK
UKHW021942290926
8980IPUK00016B/1771

9 781803 740676